TESTOSTERONE

TESTOSTERONE

AN UNAUTHORIZED BIOGRAPHY

Rebecca M. Jordan-Young

Katrina Karkazis

Harvard University Press

Cambridge, Massachusetts
London, England
2019

Library of Congress Cataloging-in-Publication Data
Names: Jordan-Young, Rebecca M., 1963– author. | Karkazis,
Katrina Alicia, 1970– author.
Title: Testosterone : an unauthorized biography / Rebecca
M. Jordan-Young, Katrina Karkazis.
Description: Cambridge, Massachusetts : Harvard University Press, 2019. |
Includes bibliographical references and index.
Identifiers: LCCN 2019012123 | ISBN 9780674725324 (hardcover : alk. paper)
Subjects: LCSH: Testosterone. | Testosterone—Public opinion. |
Masculinity in popular culture.
Classification: LCC QP572.T4 J67 2019 | DDC 612.6/1—dc23
LC record available at https://lccn.loc.gov/2019012123

For SAL, who was always our first reader

CONTENTS

TESTOSTERONE

INTRODUCTION

T TALK

I N EARLY 2017, the popular radio show and podcast *This American Life* rebroadcast an episode that host Ira Glass praised as "one of our very favorite shows." About "testosterone and just how much it determines of our fates and our personalities," the hour-long program holds within it a microcosm of testosterone's cultural meaning and power. Through the eyes of these journalists and their guests, we see the world of assumptions, assurances, confidences, and complexities this molecule invokes. The episode is like a disco ball of testosterone lore, with each tiny mirrored panel representing a bit of "knowledge" from past to present about testosterone in its myriad physical and social forms.[1]

Testosterone has a rich and varied biochemical life, and a busy but slightly more predictable social life. So familiar that it needs only a punchy nickname, T is summoned up in daily conversation and news reports in a way that most often reinforces its identity as the so-called male sex hormone, while the complexity and nuance of its many actions get short shrift. Think of the usual story about T as the authorized biography, and you'll get an idea of where we're going with this unauthorized biography. There are thousands of stories about T but surprisingly little variation. Here we're going to dwell instead on the unexpected, the confounding, the messy and fun bits. We're not writing a textbook on what T does in bodies, and we're not comprehensively reviewing research from the beginning of time or across every domain in which T has been studied.

It seemed as though every time we disclosed our book idea to someone, they automatically mentioned Glass and the *This American Life* episode. We found ourselves reacting not just to the segment but to the sense of wonder and excitement that it instilled in listeners. However eloquent and award-winning the stories, however many times the show is described as fresh and new, its disruptive strains are minor compared to the heavy thrum of T folklore that propels the narrative.

•   •   •

GLASS OPENS THE SHOW with a powerful anecdote from the episode's producer, Alex Blumberg. At the age of fifteen, Blumberg was rummaging for a book on his parents' shelves and came across Marilyn French's novel *The Women's Room,* which he recalls as focusing on a group of women who "all suffer at the hands of the various men in their lives. And there's constantly—women are slowly being driven crazy by their husbands' incessant criticism, or they're being called ugly after they get mastectomies, or they're being stifled by their husbands' emotional shallowness." The book deeply affected him, in no small part because of what else was going on in his life: puberty. Obsessed with one particular girl, he recalled seeing the "barest sliver" of her bra and experiencing an "all-consuming" desire that made him terrified that he might become like the horrible men in the novel.

From there, the episode takes a giant but seamless leap to T. "My testosterone, and how it affects me, and how I react to it, I think about on a daily basis all the time," Blumberg muses. "It often feels like there's something in my body giving me instructions that I probably shouldn't follow." A non sequitur, perhaps, but one that works because of the shopworn quality of T folklore. T is the thread that connects overwhelming desire with what we might now call "toxic masculinity"; T runs roughshod over Blumberg and other men, giving them instructions they ought not heed.

Following Blumberg, the episode turns to a man who chronicled changes to his ambition, interests, humor, the inflection of his voice, and even the quality of his speech when his body stopped making T. In his article for *GQ,* aptly titled "The Beast in Me," he wrote that "four months without the hormone taught me that testosterone . . . is everything. Everything. Not just [a man's] motivation but his very epistemology. Without the want it creates, he undergoes a gentle dry rot of body and mind, losing resolution until he becomes as negligible as a ghost." When he was treated with "copious amounts" of T, he said, "the monster took control."[2]

Glass stages the episode as a tug-of-war between rational free agency and the power of T, asking, "How much does testosterone determine?" Another answer comes from a man who experienced high T for the first time as an adult. Griffin Hansbury, a psychoanalyst in New York City specializing in gender and sexuality, "strongly identified as a woman" when he began college, but by sophomore year he "knew that I had to change my body. . . . And the only way to do that was to take testosterone." Echoing both Blumberg and the *GQ* writer, Hansbury says, "I felt like a monster a lot of the time. And it made me understand men. . . . And I would really berate myself for it."

Hansbury describes experiencing an "incredible increase in libido and change in the way that I perceived women and the way I thought about sex." Before T, he was interested in talking to women. After T, "everything I looked at, everything I touched turned to sex," to the point where even machinery could arouse him: "I would be standing at the Xerox machine, and this big, shuddering, warm, inanimate object would just drive me crazy."

Up to this point in the episode, the persona of T that's conveyed has mostly been about sex, but it takes a turn to encompass a certain kind of male intellect. Hansbury explains that after taking T, he became interested in science and understood physics "in a way I never had before," a claim that Blumberg worries "is setting us back a hundred years." It's not just intellect but also emotion that Hansbury describes as different and more masculine after T, pointing to his difficulty crying: "I'm still very much learning how to be a man in the world. There's a lot to learn." For all he still doesn't get, he is often now called "sir," and this is a victory: when he started taking testosterone, his hope was "to pass as male, to be perceived by the world as a man."

There's a guilty complicity between Blumberg, who's doing the interview, and Hansbury. Laughing, Blumberg chides Hansbury that "you've reinforced a lot of stereotypes that we've almost dispelled with." Hansbury laughs, too, and acknowledges that Blumberg is right. Their shared sentiment seems to be, *Like it or not, this is just what T does.*

Or is it? Having showcased personal narratives until this point, Glass turns to the social psychologist James Dabbs, a renowned T researcher and author of a bestselling book on T, who muddies the waters. As Glass and Dabbs talk about T's handiwork, much of their discussion echoes the authorized biography: T causes boldness, fearlessness, confidence, big muscles, and baldness. But Dabbs also seems to flatly contradict the personal stories that came before. Men, he says, "think it makes them manly and heroic and virile and sexual, which is not really true. It doesn't take much testosterone to have sex. So that's sort of beside the point."

Research gets you only so far with T's story. So Glass takes it back to personal narrative, creating a real-time testosterone drama around the question of who among the show's staff has the highest T levels. It's a homespun experiment to see if they can predict each other's T levels based on their personalities, looking first at the women, then the men.

They all agree that Julie, described as the boldest and the alpha of the group, will have the highest T. Julie quips that she expects to have the lowest T, sarcastically describing herself as "just passive me, just taking it easy. Girly, girly me. Just being feminine over here in the corner." But she fears that her colleagues may be right. When asked why she is afraid of ranking highest, she explains, "That would confirm all my worst suspicions about myself, that I'm really aggressive and pushy and sort of a hothead."

When they try to rank the men with the highest T, they consider things they never thought of for the women. How does liking sports stack up against playing sports? Balding and having muscles versus getting in fights? Jonathan jokes about the difficulties figuring out the rankings. "It's like who can yell the loudest, right? Who has the most rage? I have rage. Unfortunately, it's impotent rage. I don't know how that's going to rank." David is singled out as a tough case: "He's gay and a fan of Martha Stewart. But he's also balding and he's worked as a professional actor, which both correspond with high testosterone levels. So how do you figure that one out?" None of the women want high T; the majority of the men do.

When the results come in, Glass is incredulous. David has twice the T of anyone else in the group. Julie is indeed highest among the women, and she comments that her reading "makes me feel really bossy and aggressive." Todd, with the lowest T among the men, takes it as a blow to his manliness: "If I can't be the most manly in public radio, where the hell can I be the most manly? I kind of wish this was *SportsCenter* because then I'd be OK. . . . But in public radio?"

The Greatest Storyteller

Of all the powers of T that *This American Life* considered, here's one they missed: T is a great storyteller. Even when the T episode was rebroadcast after close to two decades, it was presented as a novel examination of persistent and entrenched notions. Rather than being anything approximating a "fresh take" on T, it's more of a flowering perennial. *This American Life* captured perfectly T's authorized biography, in which T is about libido, aggression, focus, facility with science and math. It's gendered, channeling myriad elements of masculinity.

T's familiar biography is authorized in two senses: it enjoys the social authority that comes from science, and it is also written, crafted, and narrated. This is a story that didn't just arrive from nature; it is a biography that has specific human authors. T seems to tell an inescapable truth, and the narrative sweeps away all kinds of details and smooths over contradictions. This unauthorized biography upends the notion of T as a sui generis molecular force. Like the authorized biography, this unauthorized biography is curated, but with a different aim. We turn toward the unexpected, the forgotten, the forms of evidence that are hard to absorb into T's guise as "the male sex hormone." We tell a story that's accountable to evidence, while recognizing that part of the evidence is the narrative about T. But is it even possible to separate out what T can do from what people want it to do?

•   •   •

T's STORY BEGINS decades before biochemistry enabled its chemical isolation in 1935, with scientists who were on a quest to explain sex differences using ingenious yet crude manipulations of nonhuman animals. Searching for ways to remove and then replace the "essence of masculinity," they castrated animals in their labs, recorded the effects, and looked for ways to restore the tissues and functions that were affected. For example, they implanted bits of testicular tissue, which they thought contained the substance responsible for strength, virility, and masculinity itself. They put the testicular tissue in new places, like inside the abdomen, to test the idea that the key factor was a chemical that could affect distant tissues without direct attachment. It didn't always work, but often enough the cock's comb, the bull's aggressiveness, and the rat's erection were restored, so scientists felt confident they had found the "male sex hormone." Their quest to explain sex was rounded out with parallel experiments with estrogen, thought to be the "female sex hormone."[3]

Their research was a closed loop, both grounded on and apparently justifying an understanding of the new chemicals as fundamentally about sex dualism, including expectations that "sex hormones" would be exclusive to one sex or the other, that their physiological roles would be restricted to sexual development and functions, and that they would be antagonistic. If T caused the cock's comb to swell, estrogen would make it shrivel. As early as 1920, though, scientists had reported data that they described as "surprising," "paradoxical," and "disquieting": the hormones were not sex-exclusive, and their actions were complementary rather than antagonistic. By the 1930s, researchers knew that the effects of so-called sex hormones went well beyond sex to influence processes such as bone development, heart function, and liver metabolism. Findings that contradicted the dualistic paradigm were easy enough to find: the showy feathers of the rooster were not restored by testicular implants or even by injecting T; instead, it seemed that "female hormones" were responsible for their masculine appearance. But instead of rethinking the hormone theory, endocrine researchers reclassified the physical features themselves as "neutral" rather than male- or female-typical.[4]

The research program had already leapfrogged over what would have been, scientifically, an obvious first step: meticulous documentation of everything that did and did not happen when T was removed or replaced. Researchers' fixation on sexual anatomy and reproduction meant that they gave short shrift from the beginning to the myriad effects these hormones have. A belief in sex dimorphism shaped their practices and narrowed their observations, and all of it circled back to reaffirm their beliefs in sex dimorphism. As Nelly Oudshoorn's classic history *Beyond the Natural Body: An Archaeology of Sex Hormones* elegantly demonstrated, the idea of an endocrinological sex didn't emerge from nature; it was created in the lab.

Historians and biologists have chronicled decades of similarly conceived experiments that reified the identities of T as the "male sex hormone" and estrogen as the "female sex hormone" while scientists struggled to integrate observations that didn't easily fit this paradigm. Yet the paradigm persists: everyone from researchers at the National Institutes of Health to reporters for the *New York Times* still understands T as the "male sex hormone." Our conversations with scientists also show that most are still hard pressed to describe the role of T in the healthy functioning of female bodies. In the earliest years, researchers assumed that T was produced by testes, and while they learned fairly soon that ovaries also produce T, they are still arguing about the definitive sources of T in women's bodies.[5]

•   •   •

IN FACT, T'S BIOGRAPHY stretches even further back in time. One tributary to the tale was a grand and controversial speech given at the Société de Biologie of Paris in 1889 by the French American physiologist and neurologist Charles Édouard Brown-Séquard. Brown-Séquard's talk reported the miraculous effects of what has become one of the most famous auto-experiments of all time, in which he injected himself with an elixir made of testicular extracts from dogs and guinea pigs.

What in the world possessed this renowned scientist to treat himself with this brew? In short, he was fed up with feeling old. By his early sixties, he was "so weak that I was always compelled to sit down after half an hour's work in the laboratory." First he experimented with grafting the testicles from young guinea pigs onto older male dogs in a bid to restore some of the dogs' youthful features. The experiments were mostly unsuccessful, but that didn't dampen his enthusiasm. He moved on to rejuvenating older male rabbits, and "the good effects produced in all those animals," he wrote, left him "resolved to make experiments on myself." Mixing an elixir consisting of water, blood from the testicular veins, semen, and "juice extracted from a testicle, crushed immediately after it has been taken from a dog or a guinea-pig," he injected himself ten times over a three-week period, noting "a radical change" just one day after the first injection. After three injections, he felt that his forearm strength was restored to that of three decades earlier, and both his stamina at work and his "facility of intellectual labour" had returned to prior levels. Some of the most dramatic effects may also seem, with hindsight, the most surprising. His comparative measurements showed that his "jet of urine" was 25 percent longer after the initial injection. By far the greatest effect was on "the expulsion of fecal matters," remedying "one of the most troublesome miseries of advanced life . . . the diminution of the power of defecation." He was exuberant: "Even on days of great constipation the power I long ago possessed had returned."[6]

He couldn't pinpoint whether it was the dog or the guinea pig that was responsible for the potion's punch, but "the two kinds of animals have given a liquid endowed with very great power." The improvements lasted for a month, after which time he "gradually, although rapidly," went completely back to baseline on each of his measurements—further proof, he said, of the effect of the "spermatic" fluid.

Despite what appeared to be great promise, Brown-Séquard's experiment was quickly debunked on physiological grounds as well as being criticized for building up false hopes of a fountain of youth. An editorial in the *Boston Medical and Surgical Journal* (which later became the *New England Journal of Medicine*) cautioned of a "silly season" that could encourage charlatans and quacks making mischief: "The sooner the general public, and especially septuagenarian readers of the latest sensation understand that for the physically used up and worn out there is no secret of rejuvenation, no elixir of youth, the better."[7]

In subsequent decades, however, serious researchers picked up where Brown-Séquard had left off, as if there had been no critical interruption. Top medical journals published reports of experiments covering an impressive array of techniques, subjects, and specific research aims. There were more rejuvenation experiments with implantation and grafting of testes from younger to older animals, and preparations based on the testes of goats, rams, and boars were injected into men. Testes from younger men were implanted into older men. The aims were impressively broad, from countering senility, impotence, muscular weakness, and flagging libido to "curing" homosexuality by replacing afflicted men's testes with those of "normal" men.[8]

Many of the aims and claims went well beyond what contemporary ideas about T would lead us to expect. Leo L. Stanley, chief surgeon at San Quentin Prison in Northern California for almost four decades, had explicit eugenic goals and an enormous pool of people on whom to experiment. Under Stanley's knife or supervision, more than 10,000 testicular implantations took place at San Quentin, which he claimed cured "cases of neurasthenia, senility, asthma, paralysis agitans, epilepsy, dementia precox, diabetes, locomotor ataxia, impotency, tuberculosis, paranoia, gangrene of toe, atrophied testicles, rheumatism, and . . . many other illnesses of chronic character not amenable to treatment." He was emboldened by his intervention, and said the recipients of the treatment "claim that their eyesight is improved, the appetite is increased, that there is a feeling of buoyancy, a joy of living, an increased energy, loss of tired feeling, increased mental activity and many other beneficial effects." Serge Voronoff, the great Russian surgeon who had worked with Brown-Séquard, may have made the grandest claim of all: "[The testicular matter] pours into the stream of the blood a species of vital fluid which restores the energy of all the cells, and spreads happiness." At a medical meeting in London in 1923, Voronoff announced that the Pasteur Institute's construction of an "immense park in Africa" to breed chimpanzees for their glands would place "the elixir of youth within the reach of everyone."[9]

Youth wasn't the only thing tangled up with "ideal" masculinity in relation to glands and their essences; so was whiteness. Evelynn Hammonds and Rebecca Herzig's analysis of racialization in US life sciences shows how scientists from the 1920s forward used concepts such as internal secretions, glands, and hormones to advance their interest in eugenicist "racial improvement." In 1921, Louis Berman, a physician and professor at Columbia University, published a book that, among other things, advanced a theory of white racial superiority based on racially specific hormonal balances. Thus, he said, "we are justified in putting down the white man's predominance on the planet to a greater all-around concentration in his blood of the omnipotent hormones. While the Negro is relatively subadrenal, the Mongol is relatively subthyroid. Their relative deficiency in internal secretions constitutes the essence of the White Man's burden."[10] Race has mostly gone underground in contemporary discourses about T, including research, but as we show, the entanglements of race and T are still profound.

Much research on internal secretions and hormones was conducted long before T was isolated, when it was just an idea about "the essence of masculinity." It's easy now to look askance at those claims, which often come across as outlandish, blatantly racist, or simply quackish. Today, any surgeon hawking a procedure to remedy such a vast range of ailments and conditions would be greeted with immediate skepticism not just among professional colleagues but probably by most lay people as well. Yet as much as the narratives of T have changed, a quick listen to *This American Life* suggests that one thing has remained constant: T isn't just potent, it's omnipotent. It's magic.

T Talk

"T talk" is a term we developed for the web of direct claims and indirect associations that circulate around testosterone both as a material substance and as a multivalent cultural symbol. T talk weaves folklore into science, as scientific claims about T seemingly validate cultural beliefs about the structure of masculinity and the "natural" relationship between women and men.

The root of all T talk is the sex hormone concept, whereby testosterone and estrogen are elevated as the primary hormones for males and females, respectively. With the sex hormone concept, T and its "partner," estrogen, are framed as a heteronormative pair: binary, dichotomous, and exclusive, each belonging to one sex or the other, and locked into an inevitable and natural "war of the sexes." We build upon the extensive critiques by biologists and other feminist scholars who have shown that the concept shapes how scientific information about T is gathered and interpreted, and blocks recognition and acceptance of scientific evidence that does not fit the model.[11] One indication that the sex hormone concept is still powerful is that T is constantly coded as the male sex hormone, inviting multiple inaccurate assumptions. For example, coding T as the male hormone signals that T is restricted to men and is a foreign—and potentially dangerous—substance in women's bodies, though women also produce T and require it for healthy functioning. Coding T as a sex hormone signals that T's functions are restricted to sex and sex differences, though T is required for a broad range of functions that go beyond reproductive structures and physiology.

Resting firmly on this sex hormone concept, T talk goes beyond it in several important ways. First, T lends a Stephen Colbert–like "truthiness" to a number of arguments that would otherwise appear as mere contrivances. The ubiquitous and commonsense notion of T as an overwhelming "super substance" not only substitutes for evidence but also sometimes makes any call for concrete, empirical details about what T actually does seem puzzling or obtuse. Second, while T is a synecdoche for masculinity, which itself is an abstraction, T can also symbolize biology or nature in general, as well as science and the associated values of precision and objectivity. Because T is coded as natural and in the realm of biology, just the mention of T can lend the veneer of science to simple anecdotes. Thus, by virtue of seeming to be about biology, T talk can also serve scientism—the elevation of scientific values, evidence, and authority above all others—even as it paradoxically obviates the need for evidence. Scientism equates scientific knowledge with knowledge itself, especially valorizing the natural sciences. Scientism also promotes forms of authority in which something is a "fact" or is "scientific" because a scientist says it, not because it meets any particular criterion of method. Third, stories about T are threaded through with animism: T is a willful character. When T whispers instructions in the ears of hapless men, it's clear that T has a plan, and that plan is to maintain the natural order of things. Resistance is futile.

Across the domains we examine we see T talk working both in science and at odds with it. Sometimes scientific facts fly in the face of received wisdom, while T talk fits folklore like a glove. At the same time, much scientific research on T is itself laced with T talk, with the result that researchers frame their studies in ways that anticipate familiar conclusions, overlooking (or ignoring) unexpected or contradictory nuggets of evidence within their own findings. Social elements infuse all scientific work, and in the case of T, there's a highly structured narrative that might exert more of a homogenizing effect than you see in other fields of research. Looking for the T talk is one of our core strategies for examining scientific work on testosterone. We can't peel away the T talk from the science and reveal some "pure" evidence, but it is possible to trace how T talk operates, to locate it contextually and historically in the science, and to identify the work that it does and the effect that work has. T the molecule is a fascinating substance, but T the storyteller has more power.

•   •   •

T IS AT ONCE a specific molecule and a mercurial cultural figure—a familiar villain and attractive bad boy that supplies a ready explanation for innumerable social phenomena. In an internal company memo that became an international news story, an engineer at Google, James Damore, blamed the dearth of women in tech on biology, especially a lack of testosterone. Damore's memo was important not because it represented the (fairly conventional) thinking of one particular computer engineer but because it was written and circulated at a time when Silicon Valley was under fire for having so few women in high-pay, high-prestige positions, and because it directly challenged Google's program for addressing discrimination. His subsequent firing became a cause célèbre for those who felt that the "push for equality in tech" had "gone too far," with women depicted as ruining tech by making the workplace less of a rough-edged "guy zone" where male genius can run wild.[12] Transcending the idea that T confers "engineering brain" on some people and not others, the discussion extended to broader issues of behavior in the workplace, especially the boundaries of sexual harassment.

Damore is just one in a long line of spokespeople for T as an architect of structural inequality. He followed in the footsteps, for example, of political commentator and former *New Republic* editor Andrew Sullivan, whose much-cited cover story on T for the *New York Times Magazine* in 2000 declared that T "helps explain, perhaps better than any other single factor, why inequalities between men and women remain so frustratingly resilient in public and private life."[13] That's a lot to pin on a single molecule.

There's a Jekyll-and-Hyde quality to T's reputation. T gets you to the top, for one. It's no wonder, then, that a friend in their forties takes T with the vague hope of raising their salary and being taken as seriously as their cisgender male colleagues. But the same substance that is revered as necessary for leadership, genius, and innovation can also tip behavior into the danger zone: violent, risky, aggressive, impulsive. T, so the story goes, can also get you arrested. Rounding up more than thirty studies on financial behavior, one researcher jokingly concludes that T is an "immoral molecule" that induces greed—which explains how the stock market crash of 2008 was pinned to T.[14] Several analysts advanced the idea that traders, overwhelmingly young men, took irrational risks because of high T. It's a story that sounds very similar to producer Alex Blumberg's worries about T whispering instructions in his ear that he probably shouldn't follow.

For almost any social ill or problem, it seems somebody out there is peddling the idea that T is to blame. Why are there so many more men in prison? Because T drives aggressive and antisocial behavior, so naturally men, with their higher T, get locked up more often.[15] Worried about excessive use of force by the police? The cultural historian John Hoberman argues not only that T is high among this group to begin with but that there's such widespread use of pharmaceutical T among police that sudden rages and shootings of unarmed people are a predictable, perhaps even inevitable, result.[16] What about the dog who picks fights at the dog park? We've heard T blamed even when the canine in question has been neutered. Or the ubiquity of rape in the armed forces? "Gee whiz, the hormone level created by nature sets in place the possibility for these types of things to occur," said Senator Saxby Chambliss of Georgia in a 2013 hearing on sexual assault among the troops.[17] In 2016, Geert Wilders, a far-right Dutch politician, folded immigration fears and anti-Islam diatribe into this narrative when he called migrant men "Islamic testosterone bombs" as he handed out spray cans of red paint to women to protect against sexual assault from the asylum-seekers he said make Dutch women unsafe.[18]

We've only just scratched the surface. T talk frames social issues as a matter of the chemicals functioning inside individual bodies, leaving scant room to consider power asymmetries, structural arrangements, or histories and their current material consequences. If excessive violence in American policing is explained by an epidemic of testosterone abuse among police, how does this square with the fact that people of color are so disproportionately on the receiving end of police violence? It's not a harmless theory. Likewise, if we accept the diagnosis that men dominate the tech industry because of innate capacities that flow from their higher T, then efforts to diversify tech are worse than useless: they will displace the most talented engineers. But this explanation falls apart when you look at racial disparities. If T is driving who fills tech jobs, then men of all races should be filling similar jobs, but that's not the case. And when it comes to management positions, the racial disparities are more dramatic than gender disparities, with white women moving up more easily than men of any nonwhite race.[19] There's a selectivity to the logic of T talk that is about race just as surely as it is about sex or gender. On one hand, T talk may absolve us of the very difficult work of addressing entrenched inequalities, allowing us to throw up our hands and view the current state of affairs as inevitable. On the other hand, as we show throughout the book, T and other hormones are widely understood to be malleable instead of static and fixed. Taking up new threads in endocrine research that focus on T and other hormones as responding to, instead of simply driving, social contexts and behaviors, we ask how T's malleability matters in the specific form of biologism that T talk enables.

To be clear, research on T isn't the same as popular understandings or media coverage. Research is guided by formal rules and shared commitments to transparency and precision in a way that sets it apart from other forms of claim-making, at least in its aspirations. But historical and social studies of science have established that the boundaries between scientific and popular understandings are porous, especially when the scientific work takes up questions of human personality, capacities, and behaviors. We're interested in the interface between scientific and cultural versions of T (a blurry distinction at best), keeping our attention mostly focused on contemporary scientific research, but always with an eye to the effects that claims about T have in the world. What stories about human nature does T talk support, and which do they preclude? How do group differences get explained through a prism of T, always shaped by gender, race, and class? What problems seem intractable, what solutions practical or ridiculous? How is the significance of T reconciled with other theories of power and social inequality? How does T have this much power?

Interrupting the Narrative

Echoing the philosopher Elizabeth Wilson, we take science seriously but not literally.[20] For us, taking the science seriously means that we respect science and the kind of evidence that careful, methodical, empirical investigations can provide. To be scientific ultimately means to be curious and systematic, to interrupt mental "autopilot" so that thinking can continue. Thousands of scientists over the decades have brought their curiosity to bear on T. One of our goals in this book is to bring more of their fascinating work to light, and to share in the worthy project of advancing knowledge about T by looking at studies in new ways. This is especially important because the narratives surrounding T often overwhelm the nitty-gritty evidence that's available.

When we say we don't take the science literally, we mean that we're aware that scientific findings aren't served up on a platter by Mother Nature. Instead, they are constructed out of specific research questions, the tools scientists use, and an enormous array of methodological choices, including what to measure and how, which groups or situations to compare, what statistical methods to use, and on and on. Critical excavation of science is not the same as rejecting facts, or saying that all observations or all evidence is relative. As the sociologist of science Bruno Latour observed more than a decade ago, "The question [for critical science and technology studies scholars] was never to get away from facts but closer to them, not fighting empiricism but, on the contrary, renewing empiricism."[21]

Taking science seriously but not literally thus means being ever more attentive and precise. It means not confusing anything that a scientist says or does with "science" itself, but remembering that science is about knowledge gained in particular ways. This is true even when studies have been done brilliantly and carefully. Considering context means not slipping too quickly into thinking that the relationships observed in particular studies will be true in other contexts or for other "versions" of the things being studied. With T, this is critical, because the hormone can take many forms, and they aren't all equivalent. For example, when considering the T that's naturally made by our bodies, there is total T, free T, bound and unbound T, bioavailable T, and more; there's T in blood (serum), in saliva, and in urine; there's baseline T and "reactive" T; there's T at different points in the diurnal cycle; and more. There are also quite a few versions of every single physical characteristic, behavior, or process that scientists relate to T. Thousands upon thousands of pages have been devoted to defining and refining the concept of aggression alone. So when someone claims "testosterone increases aggression," we must also ask, "*Which* testosterone increases *which* aggression in *what context?*"

Annemarie Mol, a philosopher and ethnographer of science, emphasizes how scientists tend to bracket these questions of context.[22] Following Mol, we aim to stay alert to the relevant context of evidence on T. Without conducting observations in research labs ourselves, we nonetheless looked for cues about researchers' specific research practices, such as which version of T is examined, by which measures, and in which subjects. We articulate the invisible brackets that ought to—but most often don't—appear at the end of statements about what T does. We also examine the relationships in which T becomes embedded via those technical decisions. What sort of social relations does T enable or preclude in its particular manifestations? This question is not only relevant in the context of research; T is always embedded in social relations and other matters.

For a long time, people both inside and outside of the sciences tended to think that human behaviors and biology could be broken down into nature versus nurture, each of which contributed some amount to an individual's characteristics. While plenty of people, including scientists, still hold on to this frame, a growing body of work within both the sciences and STS (science and technology studies) suggest that it's more accurate to think about biology and the social world as working synergistically rather than each separately contributing predetermined elements. Both biology and social forces are more open-ended than was previously understood. Some theorists refer to this as "entanglement," and others describe the conceptual framework as "natureculture" or use the adjective "biosocial." For instance, Columbia University recently launched a biosocial research initiative to advance the growing "realization that many biological processes, rather than being fixed, immutable mechanisms that consign people to particular life outcomes, are instead fluid, dynamic responses to features of the social and physical environments humans inhabit."[23]

An important body of work on the so-called sex hormones by the biologists Ruth Bleier and Anne Fausto-Sterling, the historians Nelly Oudshoorn and Diana Long Hall, the sociologists Adele Clarke and Celia Roberts, and others have shown that hormones are especially illustrative of nature-culture interactions. Fausto-Sterling has provided many examples of biosocial entanglements, including how and under what contexts sex/gender differences in bones emerge. Consider osteoporosis, which is generally much more prevalent in aging women than in aging men, a fact that has been overwhelmingly attributed to differences in steroids. But bone density depends on factors that include weight-bearing exercise, nutrients, sunlight exposure for vitamin D, and more. Each of these is connected to gender, and they vary dramatically across time and place, as well as across other dimensions, like social class. Fausto-Sterling points out that while boys and men often have higher bone density, sometimes the sex/gender difference is reversed, such as among ultra-Orthodox youth in Brooklyn. In this group, where scholarly activity is emphasized over physical activity, modest dress reduces vitamin D from sunlight exposure, and dietary rules that restrict dairy intake lead to less calcium, boys have "profoundly lower" bone density in the spine than girls. In much of the world, women do heavy agricultural labor, and they have denser bones. Where a large or powerful physique is seen as undesirable in a woman, women diet and are discouraged from bulking up, which can leave them with lighter and more brittle bones. Steroids including T matter, too, but they can't build bone on their own. This is a great example of what the sociologist Jennifer Fishman and colleagues call "the interrelatedness of [sex, gender, and sexuality] as material, embodied, *and* discursive sites in and through which power and power relations coalesce."[24]

This book has an important precedent in *Testosterone Rex: Myths of Sex, Science, and Society*, a witty and meticulous account by the social psychologist Cordelia Fine. "Testosterone Rex" is shorthand not for the molecule itself but for a mythical yet supposedly science-based story that anchors contemporary gender inequalities and stereotypes in evolved sex differences. Fine aims at the overarching narrative that seems to connect evolutionary theory, research on nonhuman animals, behavioral endocrinology, and neuroscience. Our work departs from Fine's by centering T itself instead of the grand narrative of evolved sex differences, and by sticking almost exclusively with human research, where we follow T into a greater range of research domains to deconstruct some especially powerful claims about human differences underwritten by T.[25]

Recently, writers across disciplines and genres have used feminist, queer, and trans theory to emphasize hormones' potential to productively transform and disrupt bodily development and functions, social relations in which hormones are implicated, and ultimately power itself. Thomas Page McBee's memoir *Amateur*, which follows his entry into the world of boxing and his ultimate bout as the first trans man to fight at Madison Square Garden, offers one of the most original descriptions of how testosterone transforms social relations. Most T research and folk narratives implicitly suggest that T does all of its work simultaneously, shifting bodies, cognition, emotion, and behavior toward more masculine expressions. McBee's account suggests that it's useful to break this down in terms of time as well as elements. T first masculinized his body, and he then had to negotiate how other people responded in socially scripted and habituated ways to his body. Likewise, though in a more theoretical and overtly political vein, philosopher Paul Preciado used his year-long autoexperiment of taking T to reflect on how T is implicated in power relations. Describing himself as "a self-appointed guinea-pig of sexual politics," Preciado ultimately advances a manifesto for using T outside of approved medical and market structures to disrupt an array of power relations including, but going beyond, gender. Whether emphasizing anxieties that circulate around hormones or exciting possibilities that the circulation of hormones introduces, the concept of malleability is a central theme in this recent work on hormones. These studies expand the emphasis on hormones as dynamic, fluid, and transitory, and this is our aim as well.[26]

Prior analyses of T or steroid research have paid scant attention to race, though both gender and sexuality—the key domains in which T is understood to operate—are always racialized. Masculinity doesn't come in just one flavor; subjective experiences, norms and expectations, and stereotypes of masculinity are all racially specific. Race isn't the only dimension that interacts with gender, of course—each of those racialized masculinities is different in a US context than elsewhere, and varies by class and other factors, too. But we found race to be an especially potent factor in

T research. Describing race's role in "the power of biology as a naturalizing discourse that has to be challenged," the historian of science Evelynn Hammonds observes that "in the United States, race serves as a dense transfer point between nature and society. It links our social structure to our individual and group biologies and it links our biological differences back to our social structure. As one of my students quipped, 'Race in America is not a biological category; it is a cosmology, an entire worldview.'"[27]

Hammonds, with coauthor Rebecca Herzig, has written one of the few accounts of how hormones have been recruited to naturalize social differences by making race seem biological. Focusing on endocrinology texts from the 1920s and 1930s, they demonstrate that some of the complex interactions between hormone biology and the sociopolitical context that we observe today were set in motion in the earliest years of endocrine research. "Just as 'genes' appear in most twenty-first century social controversies," they write, "so, too 'glands' once seemed to hold the promise of definitive answers to difficult social issues." The most important amendment that our study brings to this statement is to bring it out of the past tense. Many issues that are still tied to claims about T can be recognized among those that "gave glandular politics a particular urgency" a hundred years ago: "new demands for women's economic independence, political enfranchisement, sexual freedom, and access to education and employment." Moreover, we show a continuing "collision of endocrinopathy and race," especially in regard to scientific ideas about criminal violence.[28]

Cultural critic Beth Loffreda and poet Claudia Rankine have written that "racism often does its ugly work by not manifesting itself clearly and indisputably, and by undermining one's own ability to feel certain of exactly what forces are in play." Racial content in T research often registers on levels that elude analyses focused on the literal, intended content of studies. Thus, in addition to excavating concrete methods, we also read for implicit content, reading across different stories or studies to see how meaning is made collectively, including by resonance with broader cultural narratives or variables that are only tacit in studies.[29]

•   •   •

DELIMITING THE SCOPE OF our research is complicated, both because of the diffuse nature of discourses on T and because our own histories of following this molecule are quite long. Independently and together, we have been thinking about T for decades. Testosterone lies at the intersection of the topics covered by our first books. Jordan-Young's *Brain Storm: The Flaws in the Science of Sex Differences* upended brain organization theory, the idea that early hormone exposures, especially T, set down patterns of gender and sexuality that last for life. Via this theory, T is linked to an enormous array of psychological and behavioral traits, ranging from play behaviors to adult hobbies and occupations; gender identity and expression; cognitive capacities and personality traits; and sexual orientation, fantasies, and behavior. *Brain Storm* showed how commonsense ideas about sex, gender, and sexuality obscure fundamental incoherence in the network of studies that supposedly demonstrate brain organization in humans. Karkazis's *Fixing Sex: Intersex, Medical Authority, and Lived Experience* analyzed debates over clinicians' gender assignment and surgical decisions when they "manage" people with intersex variations. T figured prominently in these debates as clinicians drew on normative ideas about the relationships between biology, behavior, and identity, specifically the notion that T masculinizes all of these domains. Clinicians and parents rely on sociocultural norms about masculinity and femininity to make "treatment" decisions, and interventions aim to "restore" or create an appropriate binary.[30]

Our interest in T was nurtured by those projects but not bounded by them. We watched for T's appearance in scientific reports, news stories, and cultural productions from films and TV shows to advertising, as well as its surprisingly frequent mention in daily conversations we've participated in or overheard. We've been struck by how often T is on people's minds, how much anxiety it produces, and how much certainty people express about what T does.

Earlier this decade, we began to work together on examining sports regulations banning women with naturally high testosterone, according to the rationale that high T conveys an unfair "masculine advantage." The regulation was promoted as a scientific fix to the perceived problem of some women outperforming their peers. The regulation occasioned high-profile public debates about the nature of sex, gender, athletics, and more. Even in well-respected media outlets, coverage of the T regulation often relies on hackneyed ideas about testosterone as "jet fuel" for athletes and uncritically accepts policymakers' claims about testosterone as a marker of athleticism and sex. Articles often referred to these views as "the science" on T and sports, notably singular and settled. The online public comments on those popular pieces contained a sometimes hilarious, sometimes cruel jumble of scientific and folk ideas about T, such as the idea that an athlete could have "tested positive for man." We explore the science and politics of that regulation in Chapter 7.[31]

Working together on that regulation deepened the hold T had on us, and the questions we wanted to explore went well beyond the regulation of women athletes. We bring all our long-term curiosity, tracking, and analysis of T to bear on this book. Our research on T is an unbounded project that is similar to slow ethnography: there's no clear start date, and there is not an easy way to specify the boundaries of how and from where we collected our material. There is no one conference, no one journal, no one discipline, and no one domain of human life that's the epicenter of T research. To deal with this diffuse and ubiquitous molecule, we've tried to follow it into consequential places, and unexpected ones.

In recent years, scientists have begun to emphasize the reciprocal actions of steroids, including testosterone, and social context. For our research, we emphasized domains where a reciprocal influence model, known as the "social endocrinology" or "social neuroendocrinology" approach, has been applied instead of the traditional model in which steroids simply drive behavior. With the help of graduate student assistants, we conducted a network analysis of social neuroendocrinology research, initially identifying nearly 1,500 researchers. This allowed us to trace connections among researchers, institutions, and domains of analysis. We narrowed that down to researchers and groups that had produced more than a single study in the domains of interest, leaving roughly 760 researchers in thirteen research groups. In choosing domains for deeper analysis, we didn't set out to be comprehensive; doing so would mean our analysis of any domain would be necessarily thin. Instead, we looked for domains where the claims about T are especially consequential and where less critical analysis has been conducted. We selected domains where the high volume of both research activity and cross-talk between science and other social worlds would allow us to tack back and forth between them. Even this criterion allows for a huge range of potential topics. We don't examine T in clinical research except for studies of fertility treatments for women (discussed in Chapter 2) and clinical studies that pertain to T's role in athleticism (Chapter 7). In the end, we chose six main domains of testosterone research: women's reproduction, aggression, power, risk-taking, parenting, and sport. We also explore several of the big theoretical models that have guided research on T.

Among the topics that are missing from the domains we chose for analysis is how T may be implicated in developing gender identity or sexual orientation, which was a central focus of *Brain Storm*. We also have not examined practices that might be lumped under an umbrella of "taking T"—people who take T medically for sexual and other problems, athletes doping with T, or trans men and others who use T to shape gender/sex. Our primary reason for this decision is that each of these sets of overlapping practices engages claims and expectations about the sorts of changes that T will enable in both body and mind. But the subjective experience of taking T is an especially fuzzy affair because of what historian Joan Scott has called "the evidence of experience."[32] Personal experience, she observes, is often given a greater truth-value than other forms of evidence, but personal experience is just as mediated by epistemic frames and historical and sociocultural specificities as clinical or psychological studies or sociological research. One reason to bring critical scrutiny to subjective accounts of T is that we have more than a hundred years of complex cultural associations with T. Understanding the interplay of subjectivity, cultural narratives, and molecular T requires that we first identify the core expectations and experiences frequently tied to taking T, such as those related to aggression, competitiveness, and libido. Evidence of what T does in those domains must be subject to close scrutiny, and that is our project in this book. It is our hope that our deep dive into these core claims will enable more thoughtful consideration of the possibly uneasy fit between subjective experiences of taking T and evidence gathered in other ways, such as by clinical and psychological studies.

•   •   •

OUR ANALYSIS WORKS ON several levels. Within each domain, we look under the hood of studies to see how they were made, choosing high-profile researchers and frequently cited studies for close readings and methodological excavation. We don't aim for exhaustive critiques, and we don't look at the same things across all studies. Instead, we highlight methodological choices that connect in particular ways to the work that T does in the broader literature and, more importantly, out in the world. Sometimes the methods we highlight are perfectly acceptable scientific choices, but like all scientific choices, they push the process and the ultimate conclusions in specific, contingent directions. Sometimes the methodological choices we highlight are actually wrong, meaning that they violate accepted scientific practice around such issues as statistical analysis. While we are interested in the material properties of T and the great evidence that has been amassed about how it functions, that evidence is far more complex and contradictory than could be conveyed by a story of what T "really does" in our bodies and psyches. Resisting the kind of conclusive stories that are the staple of popular science writing is not always easy, because everyone, including us, would like to have some clear answers from the research. Every now and then we feel like clear answers do emerge, and we say as much. At the same time, there are always connections between scientific engagements with T and the broader cultural narratives that T enables.

We keep our eye on gaps and on evidence that is lost as well as on evidence that accumulates. Science and technology studies scholars argue that agnotology, or the study of ignorance and how it is created, sustained, and used, is as important as epistemology, the study of knowledge. Asking "What don't we know, and why don't we know it? What keeps ignorance alive, or allows it to be used as a political instrument?," scholars have examined ignorance in domains that include global climate change, military secrecy, female orgasm, environmental justice, archaeology and land claims, racial ignorance, and more. We contribute to this effort by asking not just what we know about T but also what we don't know, why, and, importantly, what facts about T are persistently lost or forgotten.[33]

The science of T is not unitary. In this biography, we show that truth claims about T are always power moves. Different forms of knowledge, including (sometimes) divergent scientific facts, ethical analyses, and "anecdata" (such as real-life experiences with T or informal, uncontrolled experiments), compete for authority. Which forms of knowledge related to T rise to the top and become authoritative resources for high-stakes debates? Examining the work T does in the world involves seeing where T is used as a foil, an explanation, and a symbol that goes far beyond and sometimes contradicts any specific evidence about this chemical. Evidence, too, doesn't stand apart but is already implicated in the idea of T, which brings evidence into being by directing research questions, focusing analyses, and sifting "meaningful" observations from apparently irrelevant ones. As a cultural idea that stands in for masculinity, T is also a transfer point or condensation point for other modes of power, especially race and class. Identifying those transfer points sometimes requires reading between the lines of a study, or across multiple studies, to see how the studies participate in discourses of power, gender, sexuality, and race.

•   •   •

As WE WERE WRITING this book, one of the most striking things for us was how often people responded with emotional force to the idea of testosterone, asking us urgent questions or making definitive statements about how it affects them in daily life. Whether it involves T in their own or others' bodies, very many people have close personal relationships with testosterone, and with the idea of T. We once attended a reception where we had been introduced as people who were writing a book about T. Over the course of the next hour, each of us was buttonholed by three separate men who wanted to talk about their own T levels—how to know if they were low, whether taking T might boost a flagging libido, whether changes in T might explain waxing and waning competitiveness. As a grand finale, as we left the reception, an attendee nearly followed us out the door as she ranted about how T had taken over her life, with two teenage sons and a husband who she felt turned her home into foreign territory. We realized we were wading into dangerous waters with this book, because people "know" about T in ways that involve a lot of shared information (right or wrong) as well as their own bodily or personal experiences. One of the difficulties is that people experience T through a hundred years of sedimented thinking. As ephemeral as ideas and talk might seem, all those soft layers adding up over time become quite hard, potentially even impenetrable. But penetrating those layers is the task we've set for ourselves. In short, we are upsetting people's relationships with T.

We hope to encourage readers to entertain ideas that challenge what they believe at the outset. We might not change your mind, but if you find yourself challenged, we hope that you will stick with us as we think it through.

Written more than a hundred years ago, T's story is a fable that has barely been updated. This book rewrites that fable.

1

MULTIPLE Ts

S EARCH THE WEB for a definition of "testosterone," and the first five
or six hits are likely to say that testosterone "occurs naturally in men
and male animals," "stimulates the development of male sex organs, secondary sexual traits, and sperm," and is "produced primarily by the testes."
These search results are more or less correct, but they are also misleading,
because they make it seem as though T is *only* made by male bodies, for
"male" traits. One would think an authoritative source such as the US
National Library of Medicine (NLM) might be more accurate in this regard, but its public education site defines testosterone as "a hormone made
mainly in the testes (part of the male reproductive system). It is needed to
develop and maintain male sex characteristics, such as facial hair, deep
voice, and muscle growth. Testosterone may also be made in the laboratory and is used to treat certain medical conditions." There are a number
of problems here, the most glaring of which is that this supposedly "male
sex hormone" is not sex-specific: it is also produced in healthy ovaries and
adrenal glands, and by conversion from peripheral tissues.[1]

The NLM's definition of T exclusively in relation to men is a stubborn
holdover from the expectations that the first endocrinologists had about
so-called sex hormones. With so much outdated information circulating,
it's no wonder that lay people tend to believe that T's presence and actions in women are negligible when even the NLM omits nearly eighty
years of accumulated knowledge about T. Some sources attempt to grapple
with this question by emphasizing discrepancies in quantity: compared
to most men, they point out, women generally produce quite small amounts

of testosterone. This focus on quantity is deceptive, though, in that it implies that small amounts of T have small effects. But dose-response curves for T aren't linear: small amounts can have big effects, especially in people who have had lower T for most of their lives. And those big effects often aren't global: for instance, many trans men find that they grow facial hair much more easily than they build muscular physiques. But clinical and behavioral studies rarely present information in a form that allows you to see the response curves that map the effects of specific doses of T onto individual participants. This is the sort of information that you glean from conversations with clinicians, and with people who have taken T.

Deeper in the search results for T, its biochemical identity as the steroid hormone $C_{19}H_{28}O_2$ starts to appear. The word "steroid" refers to a molecule with a backbone of four rings of carbon—examples include estrogen, progesterone, cortisol, and even cholesterol. This precise chemical definition is a satisfying relief to those of us who like clear answers.

But in important ways, you could also say that T as a singular chemical structure exists only in the abstract; its presence in bodies is a different story. T is found in organs of sex development, but it is also found nearly everywhere else throughout the body, including in blood, saliva, urine, the brain, muscles, skin, and the internal organs. Sometimes the molecule circulates unattached, but more often it is bound to sex hormone binding globulin (SHBG) or albumin, which are proteins in the blood. And it isn't static. Like all steroids, T is in a constant flow of generation and transformation. Some T will act directly on cells, but some will be transformed into its "downstream" steroids, estradiol (estrogen) or dihydrotestosterone.

There isn't just one testosterone: T is a multiplicity.

• • •

WE AREN'T THE FIRST to explore the concept of multiplicity. In *The Body Multiple: Ontology in Medical Practice,* Annemarie Mol showed that atherosclerosis is a different thing in a clinical consult versus a medical textbook, a pathology laboratory versus a surgical suite or an epidemiological study. Mol is not describing something like the famous parable about the blindfolded people who are told to describe an elephant, but everyone is standing around a different part of the elephant's body—it's not just that you can see different aspects of atherosclerosis in those different settings. Mol shows something more puzzling and intriguing:

what atherosclerosis is in different settings is actually at odds with what it is in other settings. Someone who complains about shortness of breath and pain while walking might be given a stress test in the doctor's office to see how well the heart can pump blood through the veins while the person walks on a treadmill. Maybe the doctor will follow up with an angiogram or a Doppler ultrasound test to gather information about the condition of certain arteries or the blood's velocity as it moves through the body. But the tests won't necessarily tell the same story. Someone might do well on the stress test but show a lot of blockage on the angiogram. Sometimes a person with very pronounced symptoms won't show very extensive blockage, and sometimes a person with no symptoms will, on autopsy, be found to have extensive blockage. This very quick scan doesn't do justice to the precision of Mol's analysis, but it gets at what we're trying to explain about testosterone.[2]

The singularity of T is an illusion, and the importance of context is profound. Here's a thought experiment: A research team is interested in the relationship between T and aggression. How would they go about studying that? Many classic experiments began by identifying a group of people who met some criteria for being aggressive and another group who did not, and comparing them to see if the "aggressive" people had higher T than the "typical" people. But there are many complexities and questions to be resolved before pursuing even this overly simplistic research design.

The first thing researchers must do is choose qualities or behaviors by which to define aggression. Though such decisions are rife with important problems and consequences, we are setting them aside here because we take them up in detail in Chapter 3. Skipping to the next step in the research process, some researchers will be interested in the possibility that T levels at some critical period in early development will have influenced aggression (this is called an organizing effect), while others will be interested in the effect of T circulating in the present (this is called an activating effect). Because in this book we focus overwhelmingly on the effects of circulating T in adults, our imaginary experiment will follow this fork in the road. Once researchers are interested in T at the present moment, they next have to decide which T to use and how to measure it.

T can be extracted from blood, muscle, or other tissues, and it is also found in urine and saliva. There are three big issues in deciding which medium to choose. First, the ease and expense of data collection are important, so researchers tend to go for what is cheapest, which is usually saliva. Second, researchers need to be able to relate their own findings to other studies, so if other studies have mostly used blood, it might not be a wise choice to switch to saliva or urine, even though either one of those may be cheaper and easier to collect. It's not easy to translate data based on one medium to another, and it's not just a matter of the scale of measurement or of the concentration in one medium versus another. T in blood isn't perfectly correlated with T in saliva or muscle or urine, and switching to another medium will affect the research results.

The third issue is the most complicated to understand. Researchers increasingly believe that the choice of which medium to use depends in part on which effects of T they are interested in. One sports scientist we interviewed focuses on T's effects in the muscle. His team had taken samples of muscle for some very small pilot studies, but owing to discomfort and expense they saw that sampling muscle would never be the technique of choice for most studies. He then wondered whether blood or saliva would be a better proxy for the T activity at the level of muscle, and found that T in saliva was the better choice. Most medical research, on the other hand, relies on blood measures; older studies often used plasma, but in recent decades clinicians and researchers use serum, which is similar to plasma but has had the clotting factors removed, which increases the accuracy of measurement.

Sometimes clinicians might measure T because they are interested in metabolic syndromes or problems with fat versus lean tissue body composition. T is related to the distribution of fat, but salivary T and blood T show different relationships to fat distribution, and there are many situations where it isn't known which is the better choice. And sometimes the major concern is pragmatic. In a study of infants looking at T as a marker of exposure to endocrine-disrupting chemicals, special gel diapers were used to collect urine—obviously preferable to researchers sticking infants with needles or even grappling with drool samples.[3]

In the aggression domain, salivary T has long been the choice among researchers. Social psychologist James Dabbs (featured in the *This American Life* episode described in the introduction to this book) used saliva mostly because it was convenient and it was easier to get large numbers of people to participate in studies if they didn't need to have blood drawn. He also believed that T in saliva, while orders of magnitude lower than T in blood, correlated well enough to be a good substitute for testosterone in blood (specifically, for free T in blood, which we discuss later in this chapter). More recently, though, some scientists who study personality and behavior have suggested that salivary T isn't just a good substitute but a better initial choice. That's because T in saliva may better reflect bioavailable T, which is the biologically active portion of the hormone. It's not a simple matter of "better" or "worse" measures, because T binds to different molecules in blood and in saliva; they are different Ts. In a 2014 evaluation of how salivary T and serum T compare, using current laboratory analysis techniques, Tom Fiers and colleagues postulated that although there are strong correlations between the two in both women and men, "sal[ivary]-T is in fact not a measure of serum free T as is often stated, but rather a separate complex matrix with its own testosterone protein-binding properties." Because of those different properties, it turns out that salivary T measures are not as accurate at lower T levels, which are typical in women.[4]

Chronology and timing are also important. If both aggression and T are considered stable traits, then when they are measured might not matter very much. But if either one, or both, is mutable, then the timing of measurement will matter a great deal. For T, timing is crucial. Fluctuations occur throughout the day, something called diurnal or circadian variation, with T tending to be highest in the morning and declining throughout the day for both women and men. It also fluctuates over a person's life course, with T levels increasing dramatically around birth, lying low during childhood, and peaking in adolescence and young adulthood. Later in life T declines through middle age and into the late years, but only in some populations. While variations across the life course and throughout a given day apply to everyone, there are additional fluctuations in menstruating women, as T rises and falls across the cycle. Major life changes like having testes or ovaries removed, whether for gender transition or cancer treatment, will dramatically alter a person's T production but not eliminate it, because T isn't just produced in the gonads.

There are seasonal variations, too, but there is no universal pattern of circadian and seasonal variations with T. And some variation is just idiosyncratic. Dr. William Crowley of the Reproductive Endocrine Unit at Massachusetts General Hospital has observed "a funny disconnect between one measurement and a later one" in a number of men he has studied: they have very low T levels at one point but later have a "perfectly normal testosterone profile." What's more, the daily fluctuations in T that have been found in US and European populations aren't found everywhere across the globe. And while many different studies around the world have found seasonal variations, the peaks and troughs don't come at consistent times, and it isn't clear what's driving them. Sunlight? Temperature? Work patterns? Variations have been observed in people whose activities have seasonal patterns, like athletes during the off-season, the training season, and the competitive season. Others, such as farmers in rural Bolivia, who have huge variations in activity across the seasons, didn't show any seasonal variation in T over the course of a year. These variations across populations, within populations, and within individuals are significant but not well understood. For most research questions, it doesn't make sense to just take one T measure and think that you have captured an individual's T as if it were a stable trait. But when taking multiple measures, it's important to try to minimize sources of variation in T that aren't relevant to the research question. Thus, most researchers measure T at the same time of day for all subjects, and for women they also factor in hormonal contraception and the menstrual cycle.[5]

Sometimes variation in T is precisely what a researcher wants to capture, such as in research on how people's T levels respond to being provoked or stressed. For some questions, such as how traders experience hormone fluctuations in response to stock market volatility, this can be studied in a real-world setting. But most researchers study dynamic changes in T by setting up laboratory situations that are calculated to stimulate a particular kind of response, like computer games with real or fictitious opponents who act provocatively. Even with these laboratory manipulations, researchers must take care to consider issues of timing that go beyond cyclic fluctuations across the day, season, or menstrual cycle. Specific behaviors, like having sex or being in a competition, can temporarily raise T levels, it's been observed, so researchers often instruct subjects to refrain from sex or strenuous physical activity during a study. But the range of activities and exposures to consider is extensive. For instance, T also responds to circumstances like being socially excluded, and chemicals such as caffeine, alcohol, and nicotine can alter T production and effects. Babies' cries, physical exertion, and sleep deprivation have all been shown to alter T levels in women and men. The various things that affect T have different time frames, but there will always be multiple circumstances, interacting substances, and life stage factors at play. Those things might have an impact on T levels, but they may also affect the person's T "responsiveness"; for instance, caffeine and nicotine intake might affect how much a baby's cry shifts someone's T production. Experimentally, it's important not to confuse the effects of, say, short-term sleep deprivation with the effects of provocation by a fictitious video game opponent or with the longer-term effects of an exercise program. Researchers have to try to take these things into account.[6]

Another timing issue has to do with the different pathways by which T can exert effects. Most descriptions of how T works say that it enters the nuclear membrane of a cell and there binds to an androgen receptor, whereby it activates target genes. That's true, but this "classical" or "genomic" pathway is only part of the story. More recently, scientists found that T also acts by binding to receptors on a cell's surface and stimulating a cascade of cellular signaling, which creates effects much more quickly than the classical pathway.[7] The timing of a T study may also need to account for whether the underlying mechanism of interest is likely to go through the slower classical pathway or the faster "non-genomic" pathway. The genomic pathway is likely to be most important for the anabolic (tissue-building) effects of T, while the non-genomic pathway is likely to be most important for emotional and behavioral effects. Thus, tracking changes in T from one day to the next might be great for studying how T affects muscle development but much too imprecise for a study of T's effects on behavior.

Once researchers have decided which bodily source of T to use, and once they decide how to address timing, they still must consider which precise chemical complex to measure. T is never simply T. Most T in the body is bound to sex hormone binding globulin, or SHBG. Some T is bound to albumin, a protein. "Total T" includes both of those bound complexes as well as T that is unbound, usually called "free T." While many researchers measure and report total T, many others think it is less informative, because such a high proportion of it is tightly bound to SHBG and therefore is not available to create reactions.[8] The latter researchers would focus instead on free T or "bioavailable T," which is the free T plus the T bound to albumin, which binds more weakly to T than does SHBG. If you're a researcher, you have to decide not only what you think is the right complex to measure but also which substance was studied in other research you want yours to relate to. If you're a reader, you have to keep your eyes wide open and distinguish the specific variations of T across studies, because they are not all looking at the same chemical complex.

Regardless of whether T is sampled from saliva, blood, urine, or muscle, and regardless of the specific chemical complex chosen for analysis, researchers must choose from among many laboratory techniques to analyze the samples. They can use immunoassays, including enzyme-linked immunosorbent assays (EIA or ELISA), radioimmunoassays (RIA), or chemiluminescence immunoassays (CLIAs). Or they can use one of the mass spectrometry methods. The choice depends upon whose T the researcher wants to measure, because immunoassay methods, especially EIA / ELISA methods, are not accurate for measuring T at the lower concentrations typically found in women and prepubescent children. But immunoassays are much cheaper and more widely available than mass spectrometry methods, so many researchers still use them. One study using commercially available immunoassay kits found readings to be off by 200 to 500 percent when compared to analysis by isotope-dilution gas chromatography–mass spectrometry (ID-GC/MS). An editorial that accompanied the study remarked that "guessing would be more accurate and additionally could provide cheaper and faster testosterone results for females—without even having to draw the patient's blood."[9]

We haven't even touched on the effects of how samples are gathered, handled, and stored (which vary widely among researchers and labs), lab choice (labs have in-house reference points and normative ranges for each steroid), or reporting units (e.g., nanograms per deciliter [ng/dL] vs. picomoles per liter [pmol/L]). Each of these choices can raise consequential and contentious issues for experts to debate.[10]

Until fairly recently, almost all scientists studied T as if it were an independent actor whose levels alone might predict some trait or behavior, usually something that is culturally coded as masculine. We explore some of the newer approaches later in the book, but many studies still focus exclusively on T levels, even though they are only part of the story. T's action in any body, whether human or other animal, male or female, depends not only on T but also on its interactions with other hormones and enzymes as well as, importantly, receptor activity. It is now common in studies of aggression or risk-taking to look at T together with cortisol, often called the stress hormone, based on the idea that cortisol modifies or moderates T's effects. Likewise, researchers who are interested in T's effects have begun to look for ways to take receptor sensitivity into account. The sensitivity of androgen receptors, which are located all over the body including in the brain, bones, muscle, fat, skin, genitals, and more, varies a lot among individuals, allowing some people to get more "bang for their buck" from the same amount of hormone.

The interaction between T and receptor activity is part of the reason T levels vary so much among different people, even people who are completely typical in terms of the characteristics that T affects. Here's an example of how that can affect data on what T does. T levels are correlated with hairiness and muscularity, because T promotes muscle and hair growth. But clinicians and researchers don't find a very strong connection between T and those things. Drs. William Crowley and Frances Hayes of the Reproductive Endocrine Unit at Massachusetts General Hospital studied T levels among healthy young men in their twenties, and while all of them were completely typical in terms of "the size of their testes . . . body hair, erectile function, sperm count, muscle mass, bone density, [and] pituitary function," 15 percent had T levels that were "more than fifty per cent below the cutoff" for normal levels in men. In an interview, Hayes "speculate[d] that some men may have highly efficient testosterone receptors—cellular traps that grab the free hormone in the blood—so that what appears to be an abnormally low testosterone level is all the hormone they need." Conversely, we spoke to a couple of clinician-researchers who used the example of healthy women with very high T to make the same point: some women might have high T levels precisely because their bodies aren't very efficient at using T.[11] Each of these examples speaks to the fact that the clinical ranges used as normative have been constructed on specific populations, and raise questions about what counts as normal. If 15 percent of men with excellent health have T below the normal range, what's the criterion for calling that range "normal"? This matters in medicine, but also in sports: how the "normal" ranges of T are calculated for men and women is central to a debate about regulating women athletes, as we describe in Chapter 7.

The complexity isn't just among individuals. The response can also vary across different tissues in the body of the same individual. One clinician we interviewed described women with very high T levels who might have a lot of facial hair but unremarkable body hair, and other women who come to see her because of dense facial hair but who don't have high T. She ascribes both situations primarily to differences in receptor distribution and activity in different parts of the body. The synergy between T and androgen receptors may also explain how baseline T may change in response to activities over the long term. In an interview with a clinician-scientist, we discussed a study showing that a prolonged course of resistance exercise raised baseline T levels in young women. He suggested that the mechanism for this change might be "receptor memory," meaning that the androgen receptor gets used to a particular T level and looks for that. The androgen receptor and T work synergistically to stimulate

both higher T levels and a greater number and activity of androgen receptors. More dramatic shifts in T, as when someone takes pharmaceutical
T or has their testes or ovaries removed, will also likely affect receptor
density and activity over the long run—but there seems to be very little
research on this.

•　　•　　•

WHEN YOU TAKE ITS material specificity seriously, T breaks into a
thousand pieces. Because T is a multiplicity instead of a singular entity,
every methodological and conceptual decision a researcher makes about
what medium to sample, when to sample, what laboratory methods to
use, and more is, in the end, a decision about which version of T will be
present in a given study. And the versions aren't interchangeable. The
range of decisions is so great that it's almost impossible to keep track of
the subtle shifts across versions of T from one study to the next. We have
only skimmed the surface of the structure of multiplicities. There are not
just more differences than we point out, but more kinds of differences.
Here, we've looked at different versions of T that coexist within one research domain but haven't considered how versions compare across domains. For instance, how do the versions of T that are typically engaged
in research on aggression relate to the versions of T that are engaged in
research on sport?

Sometimes the gaps are more readily apparent, as readers can see in
later chapters when the version of T that a researcher's hypothesis invokes is a different version than the one that is measured. But tracing the
host of different versions of T that are at play in even a single article
would require painstaking, time-consuming work. Readers would need
to track the versions of T not only within a particular study they are
reading, but backward into the literature that study builds upon, and
sideways across the studies that potentially offer—or fail to offer—
support for that study's conclusions. It would be a fascinating and worth-
while project, giving fundamental insight into the limits of how we
humans can know a world that is so dense with complexity that even
single, specific molecules have multiple personalities. But such a project
would need to stay very close to single hypotheses and fine-grained analyses, so it wouldn't have obvious bearing on the sweeping claims that are
made about T in the world.

Here, we've opted to do something different, because we felt more compelled to explore the "facts" on T with an eye toward the work that T's official life story does as it closes out its first century.

●　　●　　●

AS YOU READ AHEAD, try to keep these specificities in mind, understanding that where we have noted one gap or leap across versions, there are many others that we have not pointed out. Multiplicity is not, in the end, the main narrative that we pick up, but it is a crucial background fact. To be clear, the specificities and multiplicity of T do not mean there is no hope for a scientist who wants to pin T down. Scientists aren't oblivious to this by any means—researchers working with T are the ones who have elaborated the importance of the different versions, after all. But the specificities have a way of slipping out of view at crucial moments in studies, and especially in review articles and other synthetic statements about what "T" (singular) does. The stubborn insistence on the specific version of T that does things in the body, that relates to other aspects of bodies and reciprocally engages with behaviors, means that synthetic statements are more elusive. That may frustrate researchers who are taught that good science is characterized by simplicity. But the apparent simplicity of this molecule is an illusion.

2

OVULATION

ACCORDING TO ITS standard biography, testosterone's core function is to support male reproduction. When the story of T in the body is told chronologically, it usually begins like this: In the seventh week of fetal development, a fetus with XY chromosomes begins to take a different path than one with XX chromosomes. The gonads, initially the same regardless of chromosomes, differentiate into testes for those with the XY pattern. No sooner do the gonads become testes than those testes begin to secrete testosterone. From there, T sets off a cascade of events that send the whole organism down a male path. From the reproductive system to the brain, the fetus is now developing as a male. And the final step in the loop of reproducing maleness is that T supports sperm production in adult testes, so men have the biological equipment necessary to keep the process going down through the generations.

This is the standard version of T's role in reproduction that you'll find in every biography of T. We don't take issue with it, but instead step sideways, because there is a different story here, one that takes some sleuthing to uncover. The standard biography makes it very hard to understand what T does in female bodies. Why is it there? When it comes to female reproduction, the usual line is that T is harmful. That's not surprising, because androgens are, by definition, the hormones that generate "maleness," and the lingering concept of sex hormones suggests this will get in the way of "femaleness." But that way of framing hormones came from humans, not from nature.

DWYN HARBEN WAS HAPPILY single, but she did want to bear a child, and at almost forty-three, she knew she was running out of time if she wanted to use her own eggs. Harben made an appointment to see Dr. Norbert Gleicher at the Center for Human Reproduction (CHR) in New York City. Gleicher, a celebrated figure in infertility research and treatment, had both admirers and critics for his eagerness to push the limits of fertility medicine, including assisting women to become pregnant at later and later ages, and openly assisting prospective parents to do sex selection.

Harben didn't know whether she was fertile: she had never been pregnant, nor had she tried to be. She wasn't ready to be a parent just yet, so they agreed that Gleicher would give her ovary-stimulating drugs, harvest the eggs, do in-vitro fertilization (IVF) with as many as possible, and bank the frozen embryos until she was ready to pursue the pregnancy. It's a long and uncertain road between stimulating the ovaries and ending up with a viable pregnancy carried to term. The first step, stimulating the ovaries with hormones gauged to bring multiple eggs to maturity at once, is one of the toughest hurdles, because many women's ovaries simply don't respond to the stimulation by producing enough eggs to work with, especially as they get older. At first Harben found herself in this disappointing situation: in her words, she "flunked" her first cycle in 2003, producing only one egg in spite of receiving the maximum dose of stimulation drugs. At Harben's age, no more than 10 percent of eggs would be expected to produce embryos capable of turning into a fetus, and there's no way to tell which ones are viable without transferring them and seeing if a pregnancy is established and maintained. If her plan was going to work, she needed to produce more eggs.

Harben wasn't the sort to leave everything in the hands of her doctor, especially when so much was on the line. "Before I blow $12,000 on another cycle," she thought, "I better know whether there's something on the near horizon that looks promising." She hit the internet and came across two possible options for women who, like her, were designated as "poor responders" to ovarian stimulation. One report suggested that acupuncture might be helpful, and the other, a small study by doctors at Baylor University in Houston, Texas, suggested that treating women with dehydroepiandrosterone (DHEA), a weak androgen, before ovarian stimulation cycles might improve egg production. She began acupuncture treatment, and also began taking over-the-counter DHEA supplements. She didn't tell Gleicher, because she thought he wouldn't appreciate the fact that she was "messing with his protocol."[1]

Meanwhile, Gleicher told her that continuing treatment would be "against medical advice." He said Harben should instead come back when she was ready to get pregnant, and use eggs from a donor. Based on her own research, she understood that any ethical IVF doctor would have told her the same. But she wasn't ready to give up. She convinced Gleicher that she understood the risks and insisted on proceeding, producing three eggs and five eggs in the next two cycles. Even though there had been a small increase each time, Gleicher advised Harben that a small increase or decrease wasn't very meaningful within such a very small range. She told us, "I had to push him hard after each of the first three cycles to let me do the next one." When in the fourth cycle Harben produced seven eggs, all of which again were successfully fertilized, the upward trend was unmistakable, and Gleicher slightly modified his stance on proceeding. She was thinking that she would "go in and do [her] confession" about her self-treatment with DHEA and acupuncture after the fifth cycle, but there was a "downward hiccup" in her egg production, so she waited. After the sixth cycle, in which she produced thirteen eggs, twelve of which developed into embryos, she prepared to tell him.

By that point, Gleicher told us, he had never seen anything like it. In direct contrast to the well-established pattern that women with low ovarian reserve will produce fewer eggs over time, or at best level off, Harben's ovaries had produced more healthy eggs with each cycle. When Harben walked into his office and announced, "Dr. Gleicher, I have to tell you a secret," he was ready to listen. He read the literature she'd brought, and while he wasn't impressed with the reports on acupuncture, the DHEA study caught his attention. He said he pored over that article and was "terribly surprised." "In our training," he explained to us, "I think me, and the field generally, have mostly thought of androgens as bad for women who are trying to get pregnant. But generally accepted beliefs are not always correct." Over the course of the next couple of years, through his own clinical research and his study of the animal literature, Gleicher began to suspect that the real star of this story wasn't DHEA at all. It was testosterone.

Ovulation: The Standard Story

If you're surprised that T may play a critical role in ovulation, it's no wonder. Whether your information comes from pop science, from being an attentive biology student, or even from close reading of medical texts, it's unlikely you've ever seen T mentioned in connection with healthy ovarian function. T's relationship to ovulation is complex, and exploring it can expose how gaps in biological knowledge are established and maintained and what it takes to close them. Broader themes, like medicine's disinterest in the finer points of women's reproductive physiology, come into play, too, as do the ways that the sex hormone concept persists in obscuring certain kinds of facts.

The simplest explanation of ovulation is the release of a mature egg from the ovary, which happens roughly once a month. The standard medical version is slightly more complex, describing the development of an immature or "primordial" follicle into a primary follicle and a mature egg over a four-week cycle that mirrors the menstrual cycle. The follicles develop under the influence of three main hormones: follicle-stimulating hormone (FSH), luteinizing hormone (LH), and estradiol (E). But the ovulation story, even a streamlined version that omits everything happening outside the ovary, still has some puzzling elements.

For starters, situating the cycle in a four-week period ignores several steps that take place before this window begins. As detailed accounts acknowledge, ovaries contain follicles at varying stages of development. The follicle is a fluid-filled sac that surrounds and fosters the development of the egg, or oocyte. In order for a follicle to develop in any given cycle, it must first be "recruited" to become one of a "cohort" of follicles that together go through a stage of development that lasts several months before the final four-week cycle. Strangely, not even medical textbooks explain how a particular follicle is recruited to the cohort. To find this information, as well as other factors involved in that stage of development, you must read primary research literature.[2]

So far, this standard story of normal ovarian function doesn't include testosterone. The most active role we could find in a textbook account of T's action in ovulation is a brief mention that T stimulates estrogen production, but even that account gives more weight to T's detrimental effects, such as inhibiting FSH and causing ovarian cysts that impede fertility.[3] New research on DHEA doesn't contradict those well-documented negative effects that high levels of T can have, but it shows that this is a radically partial account of what T and other androgens do in women's fertility.

Most women have, at one point or another, been told they were "born with all the eggs they'll ever have." The image suggested is of eggs lined up on a shelf inside the ovaries, sitting undisturbed until they are tapped and released in the monthly process of ovulation. This is the classic view, in which new oocytes can't be generated after birth. But this view, which has for years acted like a lid on scientific curiosity, is wrong. The "primordial follicles" that are present from fetal life are a far cry from the oocytes (eggs) that are released at ovulation, and most "primordial follicles" will never even reach the "primary follicle" stage that is represented in the typical four-week ovarian cycle. Precisely how some of these primordial follicles develop, and which ones develop, remains some of the murkiest territory regarding reproduction. But recent studies indicate that androgens, especially DHEA and T, are crucial to the "recruitment" of primordial follicles, and to the next few stages of their development. It turns out that pushing hard against the received wisdom shifts the story's focus—and makes room for T's appearance.

More than Just a "Sex Hormone"

Sedimented ideas about the so-called sex hormones are part of what locks the standard story into place, so it's useful to back up and think about T in the bigger picture, as just one member of a big family of steroid hormones. While the first thing that comes to mind when you say "steroid" might be the anabolic steroids that athletes use to bulk up, cholesterol might be the last thing. But all the steroid hormones ultimately derive from cholesterol. In simple terms, steroids are fat-soluble substances that share a specific carbon structure and can pass through cell membranes. Steroids can act via multiple pathways: either directly, by attaching to a specific receptor that the steroid fits like a key in a lock, or indirectly, by transforming into one or more other steroids in a process called steroidogenesis. In steroidogenesis, "upstream" steroids are transformed into "downstream" steroids via dozens of possible interactions among steroids, an array of enzymes, and other cofactors (see Figure 2.1).

For the story at hand, the two most important transformations to understand involve DHEA to testosterone, and testosterone to estradiol, a potent form of estrogen. Because steroids can act either directly

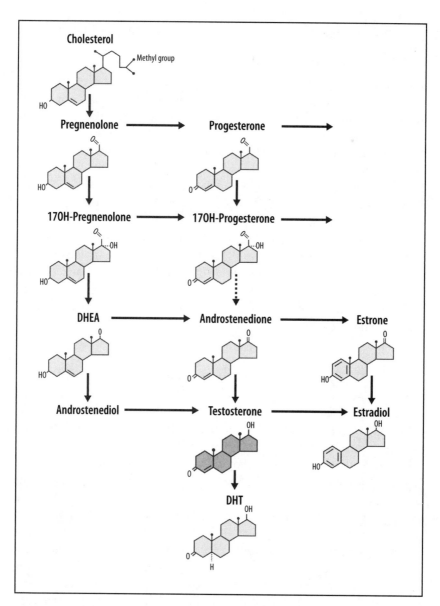

FIGURE 2.1 Main pathways of steroidogenesis.

(Created for the authors by Isabelle Lewis.)

or—through a "downstream" hormone—indirectly, it is rarely obvious which steroid hormone is ultimately responsible for any particular effect in the body. In Dwyn Harben's case, DHEA might theoretically have worked directly, or else through one of its products like testosterone or estradiol. This is where the sex hormone concept comes into play. Her body didn't care whether it was a so-called androgen or a so-called estrogen that stimulated her ovaries. But these classifications typically pose difficulties for clinician-researchers like Gleicher who want to build on results like those of Harben's self-experiment.

In the earliest days of endocrinology, before biochemistry allowed scientists to directly examine chemical structures, they had to identify steroid hormones by their effects. Researchers were keenly interested in the development of sex, which they understood to be dimorphic. So they invented the concept of "androgen," a substance that stimulated what they considered male-typical effects, and "estrogen," which stimulated what they considered female-typical effects. When a substance produced the bold coloring of a male bird, or mounting behavior in rodents, scientists called it androgen. When a substance stimulated the growth of mammary glands or sexual receptivity in rats, for example, they called it estrogen.[4]

When biochemistry entered the picture, things got more complicated. For one thing, it became clear that there were multiple substances that produced similar effects. For instance, in the 1930s, instead of finding a single androgen, researchers isolated multiple discrete steroids that had "masculinizing" effects, including androstenedione and testosterone. A more serious complication was that each specific substance didn't have only one type of action: both estradiol and testosterone, for example, have effects understood as masculinizing or feminizing. The sex hormone concept still gets in the way of seeing steroid hormones as all one closely related family, instead of as clearly demarcated "sex steroids."

As late as the 1970s, scientists puzzled over the fact that under some circumstances testosterone could stimulate "responses characteristic of sexual receptivity in castrated females of several species" and estrogen could work "like testosterone to inhibit" later sexual receptivity. These findings might have undone the classifications of steroid hormones, and we, among others, have argued that it should have. But the classification schemes were too entrenched, and in spite of its capacity to induce "feminizing" actions, testosterone is still classified as an androgen. This is a great reminder that categories are conventional, not natural.[5]

Sometimes contemporary scientists make sense of inconsistencies between classification and observations by considering the multiple possible pathways between steroids and their ultimate effects. In the 1970s, researchers devised a way to untangle some findings that they had considered paradoxical for decades, by taking advantage of what they knew about steroidogenesis. Testosterone can be metabolized into estradiol via aromatization (see Figure 2.1). Scientists at Rockefeller University discovered a way to interrupt steroidogenesis so that testosterone could not be aromatized. Decades of experiments had already established that exposure to T in early development was necessary to develop male-typical reproductive structures and physiology and, in some species, what is considered male-typical sexual behaviors. Research on aromatization showed that many of those effects were actually accomplished by estradiol. For example, if done early enough in. development, interrupting the transformation of T into estradiol prevented male animals from developing male-typical sexual behavior and also affected some physiological processes like sperm production.[6]

The long-standing habit of classifying hormones as androgens or estrogens continues to create troubles. Even now that hormones are typically classified not by their effects, as in the old days, but by the type of receptor they bind to, DHEA is a supple character. It can bind with either the androgen receptor or the estrogen receptor, but its affinity for the former is a bit stronger, so it is generally considered a "weak androgen."

How does this matter to Norbert Gleicher and Dwyn Harben? Androgen treatment for women's fertility problems went against all of Gleicher's professional training. Adding DHEA stimulated Harben's ovaries to produce eggs, but the slippery nature of DHEA meant that it wasn't clear why. Because DHEA is a "far upstream" hormone, seeing an effect from DHEA raises the question of whether it created the effect directly or through one of its products, like testosterone or estradiol. It probably wasn't the DHEA, because it is not biologically very active on its own. Some researchers say that DHEA should not even be classified as a steroid hormone but instead should be called a "prohormone"—essentially a precursor to hormones rather than a hormone in its own right. Our bodies mostly use it by converting it to the much more active testosterone and androstenedione, which can then either be used directly, via androgen receptors, or be further converted to estradiol or estrone. For Gleicher, the big question was which of DHEA's downstream products was the active substance.

The Baylor study that first drew Harben's attention involved giving five women DHEA. After DHEA, they were treated with FSH to nudge the immature eggs into maturity and release from the ovaries. The women hadn't previously responded to FSH stimulation, which is why they were classified as having "low ovarian reserve." But after two cycles that included DHEA priming, their ovaries produced viable eggs, and one of the women got pregnant and delivered twins as a result. The researchers were unsure how to explain their results, but they seemed to lean toward the explanation that DHEA had increased estradiol in the ovarian follicles. In short, their idea was that increasing estrogen boosted ovulation. They skipped right over the T.[7]

That T of the Month

In late 2016 we traveled to Gleicher's offices overlooking Park Avenue in Manhattan to talk about T and ovulation. The authors of the Baylor report that Harben brought to his attention, he said, didn't focus on T; they briefly noted that the women's T levels more than doubled after stimulation, but there's nothing about T in the discussion section, where they interpret their results.

Gleicher picked up that T levels had doubled, and made sense of this by turning to a different timeline for the ovarian cycle. The blood levels of hormones at the point of ovulation don't offer information about how the follicles get to the point of development where they are ready for "recruitment" in the last two weeks before ovulation. So rather than focus on the moment of ovulation, Gleicher's account moves backward in time to the stage of follicular development where the follicles are sensitive to different substances, in particular to T.

To put this picture together, he went to "the peripheral literature. That is often where the most interesting ideas are coming from," he said. He found that researchers had been investigating the role of androgens in ovulation among nonhuman animals for more than a decade. These studies showed that androgens like DHEA and T played several positive roles in the early stages of follicular development, especially in mice. This didn't square with the dominant medical view that androgens are "bad for women trying to get pregnant," he said, but in addition to the Baylor study, he was faced with his own androgen-treated patient who didn't suffer any ill effects and whose pattern of egg production was unprecedented in his many years of practice.

Bolstered by a plausible theory, a couple of promising case studies, and a general willingness to make bold moves (a trait that has earned him both accolades and a great deal of criticism), Gleicher and his colleague David Barad moved quickly to begin using DHEA with other patients in their fertility practice who had not responded well to the usual IVF regimen. It didn't work for everyone. By closely monitoring levels of an array of steroids after DHEA treatment, Gleicher and colleagues determined that some women who don't respond to DHEA have a problem with converting DHEA to T. So in addition to using DHEA, they began to test women for the enzyme that converts DHEA to T. If it is low, they treat the women with low doses of T, and he says they have seen success with that, too.

It would be hard to overstate what a surprise this is in the typical clinical view of androgens and women's fertility. Until here, we've skirted the vast research literature on T and ovarian function. But that literature is all about pathology, especially the role of androgens in polycystic ovarian syndrome (PCOS). PCOS is a diverse syndrome, but many women with PCOS have fertility problems because their eggs don't proceed through the full sequence of developmental stages necessary for ovulation. T is usually but not always elevated in PCOS, but high T is one of the first things clinicians look for to diagnose PCOS, and lowering T is one of their common treatment targets. When we ask clinicians about the role of T in women's reproduction during interviews and conversations, not only is PCOS the first thing they mention, it is consistently the only thing. The positive, possibly obligatory role of androgens in normal ovarian function has so far not reached most practicing clinicians, at least outside of fertility specialists.

Prepared to go against the grain, Gleicher and his colleagues, especially David Barad, pushed ahead. With a third colleague, Andrea Weghofer, they wrote a 2011 article that addressed the paradox between their clinical experience and the logic that was ingrained in their training as endocrinology and fertility specialists. Posing the rhetorical question of whether androgens were "friend or foe of fertility treatment," they noted that the answer is neither. Instead, there is an optimal level, a sweet spot where androgens are critical to ovulation. This mind-bender emerged in the context of treating women with fertility problems, but it holds for ovulation across the board. A major 2015 review of both human and nonhuman studies pronounced that "optimal levels of androgens are required for normal ovarian function." This corroborates Gleicher and colleagues' hypothesis that the question is not whether androgens are "friend or foe" to female fertility but how much is best.[8]

High T interferes with ovulation in two ways. It interrupts the last stage of follicle development and suppresses the surge of luteinizing hormone, which is the trigger for ovulation. But very low T turns out to be a problem, too. T is beneficial, perhaps even necessary, for the earlier stages of follicle development. The crucial new information is that it's not because T is converted to estradiol: rather, T has direct positive effects on ovulation. Animal studies have given the clearest evidence for this. One key research approach has been to interrupt steroidogenesis at the point where T converts to estradiol. In animal models, if the enzyme necessary for converting T to estrogen is blocked, DHEA still has the same effect on follicle development. Recently, researchers have developed another model, in which they disable androgen receptors in animals including rodents, sheep, and primates. Knocking out the androgen receptors creates serious fertility problems, including premature ovarian failure.[9]

While there have been a number of studies of DHEA and T, only one randomized controlled study in humans, where women with similar profiles undertake ovarian stimulation with or without initial DHEA treatment, has been undertaken so far. While the DHEA group had six pregnancies to just one in the placebo group, it's a single study of only thirty-three women.[10]

More studies like this might be hard to come by. As Gleicher pointed out, many of the fertility patients he and his colleagues see are running up against the limits of their age, and with such a short window for achieving pregnancy with their own eggs, the possibility of getting a placebo in a randomized trial is an unacceptable gamble to them. Another reason could be even more important: DHEA is readily available, and word began to get around years ago that it might boost the chances of becoming pregnant. When the sole randomized trial was trying to recruit participants in 2008–2009, thirteen of the sixty women they initially identified as eligible had already been taking DHEA before the study team could recruit them. Previous experience with DHEA makes someone ineligible for a randomized study.

This brings us full circle. By Gleicher's account, there was initially "some resistance" in his field to the idea of using DHEA, but now there is "a rapidly growing consensus" that it works. Fertility specialists have come to accept that androgens play a major role in the early stages of follicle maturation. A recent survey of IVF clinics internationally found that about a third of them are now using DHEA stimulation, and Gleicher believes that the rate of use has "continued to grow dramatically."[11]

But some experts think the hype has jumped ahead of the evidence. Writing in the journal *Human Reproduction,* Kayhan Yakin and Bulent Urman note that DHEA "supplementation has been hailed in the IVF world" as a "miracle drug," but they caution that "it is time to reconsider whether clinicians should be offering DHEA to their patients based on reliable scientific evidence or whether to regard it as an empirical drug with possible but no proved benefit." Likewise, a major 2015 review acknowledged that treatment with DHEA or T seems to modestly improve live birth rates for women identified as "poor responders" to ovarian stimulation, but it suggested that the quality of the evidence, at this point, is only moderate.[12]

There's still so much that remains unknown. Is T what endocrinologists call "obligatory" in ovulation, meaning that ovulation cannot occur if T is not present? Or is it what they call "permissive," meaning that T improves ovulatory function by allowing other hormones to have their full effects? One reason we don't yet know is that the research on this question is barely in its infancy. That may be because T is still understood as the "male sex hormone."

Enacting Testosterone

When T is discussed in relation to reproduction, it is nearly always about development of the male reproductive system or spermatogenesis. This terrain is changing even as we write. When we were first drafting this, a search for a literature on T's role in female reproductive functions turned up only problems, such as polycystic ovaries, irregular menstruation, and infertility. To find anything positive required a long and hard search, and enough comfort with biochemistry to ferret out the pertinent information. Now, though, a web search for "testosterone and female fertility" yields a number of results that highlight T's positive role. None of these widely available sources goes quite as far as saying that T is crucial to ovulation, but that seems to be the case. Here's a material action of T that nobody expected, is simultaneously explored and resisted, and remains contested.

The story of T and ovulation is in part a story about the difference between clinical research and basic research, but it serves as a useful example of why it's so hard to make definitive statements about what T does. T isn't just out there doing its business on its own. You have to look upstream to DHEA and downstream to estradiol to consider the possibilities. Figuring it out in one context, even experimentally, doesn't mean you can declare for sure how it works in another. Philosopher and ethnographer of science Annemarie Mol uses the term "enactment" to explain how things come into being through specific practices. A different context means that you are looking at a different object.[13]

What does the concept of "enactment" mean when the object in question is so apparently singular and material as the molecule testosterone? First, it is perhaps obvious with an object like testosterone that it is not entirely discursive: there is a material object that exists outside of how we know, understand, or study it. Its materiality lends the illusion of constancy and singularity. While phenomena like gender and sexuality are, on the one hand, profoundly naturalized, no one believes that they can isolate gender or sexuality under a microscope. Testosterone, on the other hand, seems like something you could preserve in amber: singular and immutable, you could move it from one place to the next with no real consequence for the fossil inside the amber. Indeed, the static ball-and-stick models by which many people learn about biochemistry encourage understanding each molecule as having a permanent "essence" independent of its context. The point here is that we have no access to the object T except through our specific engagements with it: we use particular tools, ask specific questions, gather some kinds of information about it while letting other information fall to the side. Importantly, the enactments of T aren't just differences in the way that human actors (scientists, patients, journalists) approach the question, but the specific material circumstances in which the molecule engages with other material actors in the body. As we described in the discussion of multiplicities, it matters a great deal whether testosterone is bound to albumin or some other substance, is "free," is circulating in the blood, is in the saliva, and so on.

Gleicher's account doesn't show that what we knew about ovulation before was "wrong." Shifting the timeline doesn't just shift the optics through which we can see the relevant actors in ovulation; it changes the actors involved. It's helpful to remember that the objects in question are organic, and that developmental processes mean that in some ways, there is no constant object to keep your eye on. The developing follicle is at some stages sensitive to the effects of FSH, but at an earlier point in time it does not have that sensitivity. This is a material object with real properties and affinities, but it is not a constant. The context is always crucial, and involves time as well as space.

Is there room for social context, as well as time and space contexts? Yes. The T that we know emerges or disappears in the context of the specific tools that are used to study it. The timelines are examples of this, as are the decisions about which hormones to measure in the first place, and which to comment upon when the discussion section of an article comes along. Which elements of the problem drop out as not meaningful or interesting? Which elements rise to the surface? The aim of this question is not to settle an imagined dispute about who is right. Different ways of studying an object don't show different sides of it; rather, they enact different versions of it. It's not a matter of trading one version for another, or of alternating between them with the hope that one will eventually be revealed as the real version.

At the same time, some versions preclude certain questions and align more easily with dominant ways of knowing the body. Some accounts of ovulation seem more obvious and even plausible because they involve familiar actors; they fit a repeated script. The Baylor team's assumption that DHEA boosted ovulation through estrogens was predictable because of the historically entrenched but scientifically wobbly idea that "male" hormones support "male" characteristics and "female" hormones support "female" characteristics. This same logic explains why there is still such a dearth of literature on the role of androgens in ovulation, in spite of the fact that evidence has converged over the last twenty years to confirm that T and other androgens play a key part.

Scientists seem to be having a hard time absorbing the fact that this "male" hormone is crucial for the definitive female function, ovulation. We asked Gleicher whether thinking of testosterone as the "male sex hormone" might have slowed down the understanding of the role that testosterone plays in ovulation. He hesitated, saying that he didn't want to be "too PC" about the issue, but eventually concurred. "Yes, I suppose it probably did," he said. "Of course men have much higher testosterone than women, but it is probably equally important in men and women." In general, scientists are supposed to observe the principle of parsimony—the best explanation is the one with the fewest variables. According to that principle, it would make the most sense to attribute the effects of DHEA to T, as T is synthesized more directly and in more abundance from DHEA than is estradiol. But the general prejudice against linking T to a core "female" function raises the bar for the evidence required to accept this link.

The Baylor team's version of the ovulation story not only squares with the sex hormone concept but is consistent with accounts of ovulation in textbooks and popular media. Only the last two weeks of the egg's development are detailed. Everything else is squeezed into a single sentence, or maybe two, that mostly describe how the stages between the eggs' appearance in the fetus and their "recruitment" for the final stage before ovulation are unknown. The gap is made unimportant by the familiar story that "women are born with all the eggs they'll ever make" and by a long habit of neglecting to research the details of ovulation.

Our ovulation stories demonstrate the utility of pragmatic epistemology: what is the right science for the job at hand? For Harben's purposes, opening up the black box of the follicle's development in between the primordial stage and the recruitment of primary follicles was crucial: otherwise, she simply had "low ovarian reserve" and had no way of making more mature eggs. The push to understand her situation came initially from her own desire to make something happen, rather than a deliberate inquiry or demand to "know." Further elaboration came from Gleicher's pragmatic inquiry into how he might replicate the results that Harben got. Harben's self-treatment opened a potentially lucrative new market for Gleicher and the field as a whole (though he didn't say as much). An experienced and savvy businesswoman, Harben recognized this and negotiated a partnership with Gleicher and his corporation; she co-holds the patent on the DHEA treatment that they use now at the clinic. All the tricks that fertility specialists had in the bag at that point didn't include anything for women with severely diminished ovarian reserve; if hormone stimulation didn't produce mature follicles, the women were never going to ovulate. They just had to accept that they weren't getting pregnant with their own eggs.

It is helpful also to take an agnotology perspective—that is, to think not only about what has been persistently unknown or forgotten about female reproduction, but also about why these elements are forgotten, and how precisely that forgetting happens. Our examination of ovulation suggests that the sex hormone concept is a powerful mechanism of ignorance, encouraging some questions and foreclosing others, and blinding researchers to certain data that they have long had in hand. Another mechanism of ignorance that keeps T out of view has to do with the long-standing characterization of female development as passive, an oxymoronic trope that many feminist STS scholars have identified. From a biological point of view, this is implausible on its face, because all development is by definition active, engaging specific material elements and time frames. Accepting that female development is passive underlies the idea that eggs simply sit in the ovary until they are released, keeping scientists' curious eyes away from the events that prepare eggs for that stage, and truncating the timeline of the ovarian cycle.[14]

The force that this model exerts is strong. Even with critical awareness that the concept of female developmental passivity blocks knowledge, it was not immediately obvious to us how we could incorporate the knowledge about T's role in ovulation. We showed Gleicher a standard chart of the ovulation cycle and asked, "How would we incorporate T? Where would it go on this chart?" He told us: "You would need a whole new chart, because your time frame here is all wrong." That is, *all wrong for T*. The chart doesn't misrepresent the last weeks of follicle development, but T's actions come before that. T stimulates follicle development and prevents premature oocyte death in the earlier stages, roughly the two months preceding our original chart. Figure 2.2 is a revised chart. The left-most column shows the period in which T is important; to its right, we have mapped the standard version of the ovulatory cycle, which includes menstruation, ovulation, and the decay of an unfertilized egg, and features FSH, LH, estrogen, and progesterone, but not T.

Another mechanism of ignorance is the tendency for science to operate in silos—that is, different scientific practices don't always engage one another. For example, even as some practices in fertility research and treatment enact a version of T that is crucial to female fertility, others foreclose T as a meaningful or positive contributor to female functions. The research that allows T to have a role in female fertility has been going on for a couple of decades now, and the idea has gotten a lot of traction among fertility specialists. But if you look elsewhere, or even look into the most dominant and authoritative stories about female fertility (such as those found in medical textbooks), the descriptions of T as supporting "male" structures and functions is as stark as it ever was.

The story of ovulation shows that the trouble with T isn't just cultural. The case of T's role in ovulation, and how slowly the knowledge of that process is created and disseminated, is an important example of how T's social identity is consequential to what we know about T in the most formal, material, scientific sense. The recent studies on ovulation have the

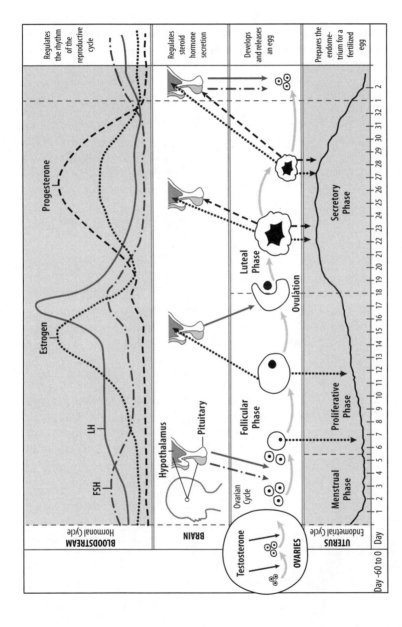

FIGURE 2.2 Classic depiction of the ovulation cycle, showing follicular development in relation to brain, uterus, and key hormones, updated to include testosterone's role. (Created for the authors by Isabelle Lewis.)

potential to upend a century of understandings about what testosterone and its fellow androgens do. Indeed, playing a central role in ovulation undermines their very classification as androgens. And that's one of our main reasons for going into the details of these studies: it's a great case study for examining why it's so hard to learn new things about T.

3

VIOLENCE

I N A MILITARY courtroom at Ft. Benning, Georgia, on March 15, 1971, prosecutors made their closing arguments against Lt. William L. Calley for the most notorious massacre of the Vietnam War. Three years earlier, dozens of US soldiers had murdered, raped and/or mutilated scores of unarmed civilians of all ages in the Vietnamese hamlet of My Lai. Though more than a hundred men participated in the atrocities, Calley was the only person ever convicted.

The brutality and scale of the violence at My Lai were beyond belief to most Americans at the time. But an explanation of sorts could be found on the front page of the *Washington Post*: nestled alongside an article reporting Calley's trial was a second article proclaiming, "Army psychiatrists are studying the relationship between male sex hormones and aggression to find a way to keep irrational killers out of the military." No one was directly saying that this particular war crime could be chalked up to a ferocious case of testosterone poisoning, but the juxtaposition of the articles, and the quotes from research psychiatrists about their T studies, undoubtedly alluded to a connection between T and the massacre. "We're trying to weed out people who can't handle their aggressions— people who are so aggressive that they haven't learned how to control it," said Dr. Robert Rose, one of the psychiatrists whose work was featured in the report.[1]

One of the most familiar and enduring stories in testosterone's authorized biography is that it drives violent aggression. This association didn't start with the My Lai massacre. In their examination of concepts of race in US medical and scientific thinking, the historians of science Evelynn Hammonds and Rebecca Herzig show that mid-twentieth-century experts considered endocrine disorders, also known as "endocrinopathies," to be an important cause of crime. Nevertheless, we began with the *Washington Post*'s pairing of Calley's trial and Rose's research because it is such a stark example of how scientific facts are, to quote the anthropologist Amade M'charek, "less about discovery and more about the making of reality by assembling heterogeneous material." And because it's a great example of how scientific facts, once established, are so difficult to dislodge. The study itself, not to mention the wildly misleading association of the study with the massacre at My Lai, would be unlikely to make it into print today. Yet the study's central finding—that high T levels are related to "more serious offending" or "more violent crime"—has shed its embarrassing origins and lives on as a simple empirical truth, no longer weighed down by the specifics of a study that doesn't match the methods or theories of contemporary behavioral endocrinology.[2]

The notion that T drives violent crime is like a zombie, a fact that seemingly can't be killed with new research or even new models that would make old research irrelevant or subject to new interpretations. In this chapter, we aim to show both how this zombie fact was assembled and, even more importantly, the reasons it won't die. This requires excavation not just of methods but of dated studies that are badly flawed and should have died out, yet live on in the literature; they've never been closely examined, and they have a strong afterlife. Because it is so widely accepted, this zombie fact shapes understandings of criminal violence as a matter of individual or group biologies, constraining the remedies we can pursue or even imagine.

We organize the excavation of the "fact" that T drives criminal violence around key works by Robert Rose, James Dabbs, and Allan Mazur, three researchers whose work has been central to the construction of this scientific fact. The works we examine are the go-to studies that researchers still use to validate the claim that high T is linked to extreme aggression or criminal violence; there are no sustained critiques in the literature for any of them. In fact, when we recently searched the web for popular beliefs about T and aggression or violent crime, the first few results are classic studies of T in prisoners done by Dabbs and colleagues in the 1990s, work by Mazur and his frequent collaborator and fellow sociologist Alan Booth, and an article dedicated to Dabbs's memory. Just below these studies is a 2012 *Scientific American* article that repeats the claim that "regardless of their gender, the most violent prisoners have higher levels of testosterone than their less violent peers"—a claim that has become so familiar in the popular science literature that it no longer needs a reference to particular studies. These and other popular retellings casually affirm a link between T and criminal aggression, even as they use the contemporary hedges that make the story seem scientific and plausible. No longer directly "causing" aggression, testosterone is "necessary for violence, but . . . not, on its own, sufficient." Criminality is a key element of T's classic biography, even if T isn't the renegade actor that William Calley was made out to be. These days, in what we must take as the straight-faced words of science journalist Christopher Mims, "testosterone is less a perpetrator and more an accomplice—one that's sometimes not too far from the scene of the crime."[3]

In this chapter we also look at the interplay between studies of T and aggression, and concurrent public discourse on violence as a social problem, especially the class and race politics of violence. From the beginning, the science and scientific narratives that link T to aggression have been about both individual bodies and broad social problems or trends: Why are men violently aggressive more often than women? How to weed out the overly aggressive soldier? How to know who is destined to be a violent recidivist, versus a person who could be rehabilitated? These aren't mere matters of scientific interest; they are political as well. Testosterone has figured in discussions of one of the most explosive issues of our times: the highly racialized interplay between violent crime and the use of excessive police force against unarmed civilians, especially black men and other men and women of color. In a neatly drawn parallel, high T is seen as driving both violent crime and the rampant overuse of force in policing. For two decades, Allan Mazur, a sociologist and elected fellow of the American Association for the Advancement of Science, has claimed that high T levels in young black men might explain broad social patterns of crime, such as the FBI statistics for the United States in 2013 that show "38% of murderers were known to be black . . . and 51% of victims were black." That same year Charles Ramsey, then police commissioner for the city of Philadelphia, said that testosterone abuse was responsible for "some of the worst domestic violence" among police officers and made officers "more aggressive when interacting with citizens," calling out what historian John Hoberman has called a "'steroidal' policing style."[4]

*   *   *

WE DON'T SCRUTINIZE every aspect of every study—exhaustive forensic examination is not necessary to see how the widely accepted idea that T drives human violence is, at the very least, grossly overblown. The very same researchers who have helped to establish the fact status of the connection between T and human aggression have acknowledged for decades that this link is "tenuous," "weak and inconsistent," "inconclusive," and "elusive."[5]

As we were writing this book, we continually surprised people by telling them that scientific studies on this point are decidedly mixed, made up of evidence that is more contradictory than supportive. The idea of aggression as an endocrine problem dates to the earliest research on hormones, but the foundational studies on T and human aggression date to a roughly thirty-year period from the 1970s through the 1990s. These studies are where you will find grand claims that higher T levels have been found in violent criminals and others whose behavior is "antisocial" and specifically aggressive. Later studies, especially lab research measuring subtle aggressive actions like "punishing" another player in a computer game, are layered upon these foundations in a way that seems to extend the connection between T and outright violence, suggesting that even relatively minor hostilities can be traced to the effect of this hormone.

And yet gold standard studies consistently have found T to have no effect on aggressive behavior or feelings: these are double-blind, placebo-controlled studies in which neither the investigator nor participants know who is getting T versus an inert substance. A number of such studies have involved raising T to supraphysiological levels (that is, well beyond the upper limit of the range seen in healthy men), and have also included asking the men themselves as well as significant others to report on the men's mood and behavior. No effect of T on aggression, anger, or hostility has been observed in the studies that have used this design. But at present, researchers still read the overwhelmingly negative evidence on T and human aggression as "weak and inconsistent," seeing in that an intriguing doorway to new knowledge rather than an indication of a dead-end street.[6]

One of the most fundamental questions in research on T and aggression is what counts as aggression, and whether "aggression" is even a useful construct for thinking about how T might shape human relations of hierarchy and conflict. Researchers' arguments about how to measure aggression and how it is related to other constructs like "violence" and "dominance" are important in assessing the evidence for T's effects, and we will look at some of those discussions. In this chapter, we follow the trajectory of one thread of this research, which explores the link between T and criminal and antisocial behavior. These studies are especially important both in the lay imagination and because they seem to have what scientists call "external validity," meaning that the way the variable is measured in the study is connected to a real-world phenomenon. Consider the contrast between a typical laboratory study of aggression and a study of "criminal violence": "retaliating" against another player in a computer game or making unfair bargaining offers can be interpreted as aggression, but these are the kind of small, everyday affronts that anyone might exhibit on a given day or under particular circumstances. Committing a violent crime, on the other hand, or being rated by people in your life as consistently hostile in the real world may be more than just a quantitative difference in negative behavior or affect. Criminal and antisocial phenomena are matters of problematic personality types and social problems, not just minor variations in behavior. Unlike lab studies of minor aggression, studies that link T to violent crime and antisocial behavior suggest that social problems of violence are rooted in individual biologies.[7]

The "Mulder Effect"

At the time his T studies were paired in the *Washington Post* with the story of My Lai, Rose was a young civilian psychiatrist at Walter Reed Army Medical Center. He would go on to become a superstar in psychosomatic medicine, a new interdisciplinary field that concerned the complicated and reciprocal relationships between physiology and psychology. Though most of Rose's T research on human aggression was conducted together with Major Leo Kreuz, a psychiatrist at Walter Reed Army Hospital, Rose had an especially high profile because of the extensive research on T in primates that he conducted before and during his human studies. While he hasn't worked on T in decades, his work with soldiers and prisoners helped to shape the field and still anchors the idea that high-T men are more violent.

When the *Washington Post* previewed Rose's research, he and Kreuz had just completed a landmark study on prisoners at the Patuxent

Institution in Jessup, Maryland. Right around the time the study was published, it was also covered in the *New York Times,* where the question of how to measure aggression took center stage. Because many of the ultimate findings would counter existing research on T and aggression, both researchers and journalists wondered if the discrepancy might point to the problem of how to measure aggression. As the *Times* reporter put it, "Is it observable outward acts . . . or is it negative aggressive *feelings?,*" noting that one researcher had identified at least nine different kinds of aggressive behavior, including fights over territory, fear-based aggression, and males fighting over females. Jostling over which measure was best for studies of humans was more than an abstract debate; saying that the researchers must have just used a bad measure was also a way to salvage the idea that T is linked with aggression even when studies failed to support that idea.[8]

In retrospect, given these measurement concerns, Kreuz and Rose's choice of the maximum-security prison at Patuxent may seem ideal. Studies of prisoners had always held a special place in the literature on T and aggression, as being convicted for violent crime is often seen as an especially valid measure of real-world aggression, both more meaningful and more objective than measures based on self-report, personality assessments, or behavior in contrived laboratory situations. To Rose and Kreuz, this maximum-security prison may have seemed like the ideal place to research the relationship between T and criminal violence, as it had been founded to house Maryland's most dangerous criminal offenders. When Kreuz and Rose did their study, inmates were sentenced to Patuxent under the "Defective Delinquent Statute," a law that was unique to Maryland and allowed indefinite confinement of "habitual criminals considered to be a clear danger to society for psychiatric treatment at this institution." Working inside the prison meant Rose and Kreuz would have access to men with especially violent histories. Moreover, because Patuxent was under a legal mandate to keep highly detailed records, the researchers would have access to fine-grained information about each man's criminal history and behavior while in prison.[9]

Kreuz and Rose set out to test the idea that aggressive prisoners would have higher T levels than nonaggressive prisoners. They selected twenty-one inmates and divided them into two groups, "fighters" and "nonfighters." "Nonfighters" were reported to have been in one or no prison fights; "fighters" had to have been in two or more. Three different aggression scores were calculated for each inmate. The first score was culled from prison records and included items you might expect, such as physical fights, making threats, or destroying property, but also some items that you might not expect in an inventory of violent behavior, such as cursing or refusing to obey officers. The second set of data came from three standardized, written psychological tests to assess things like subjective feelings of aggressiveness. Finally, they looked at past criminal offenses, including the type of offense, frequency of offending, and age at each offense. All told, they examined close to two dozen behavior measures in the sample of just twenty-one men. And then there was the testosterone. To make sure they got reliable T measures, they drew each man's blood multiple times, always first thing in the morning to control for T's diurnal rhythm. This was considered top-notch research at the time: they were being very thorough, and if there was a relationship between T and aggression, they were bound to find it.[10]

Here's the kicker: despite hundreds of citations that characterize this study as proving a link between T and aggression, the study actually undermines that link. How is this possible? Unsurprisingly, the number of reported fights a given prisoner had correlated to a broad pattern of aggressive behavior while in prison. But T remained stubbornly innocuous. T levels did not predict whether someone was in the "fighter" category, displayed additional aggressive behavior in prison, or scored higher on psychological scales of aggression. When their original plan left them empty-handed, Kreuz and Rose examined the men's records of past convictions for crimes of physical violence like assault and murder. Again, they found no relationship with T. They then cut the data yet another way, examining the kinds of crimes for which men had been convicted before age nineteen, some of which had likely occurred over two decades earlier.

With that last move their persistence paid off. The men with what they called "more violent and aggressive offenses during adolescence" had significantly higher T levels than men without such adolescent offenses. They also came at this link from the other direction, starting with T levels. The five men with the highest T levels were all categorized as having committed "violent and aggressive" crimes in adolescence. None of the five men with the lowest T scores had this kind of adolescent offense. To drive the contrast home, the researchers showed that all the men with the most extreme T scores, whether high or low, had adolescent convictions for the (implicitly nonviolent) crimes of larceny or burglary; it was only the "more aggressive or violent crimes" during adolescence that were associated with T.[11]

How did Rose and Kreuz get to the link between aggression and T? Hardly a straightforward test of a hypothesis, this was more a chronicle of how to repeatedly massage data until it yields the desired result. They wanted to believe. We call this "the Mulder effect." Fox Mulder is an FBI agent on the long-running and newly re-made TV program *The X-Files*. "I want to believe" is the phrase on the now-iconic UFO poster behind his desk. In this science fiction drama, Mulder and his partner, Dana Scully, investigate unsolved cases with paranormal elements, with a heavy emphasis on aliens. In the series, the role of FBI investigator is more like that of a scientist who searches meticulously through evidence with the aim of uncovering the truth. But for Mulder, like all scientists, objectivity is elusive. As media critic Laura Bradley describes, "Wanting to believe is Mulder's core vulnerability. And for Mulder, wanting to believe is different from blindly believing. He is still an FBI investigator, of course, always trying to examine the facts—even if he occasionally searches for facts to support his theories instead of theories to support the facts." As we read the studies on aggression and other characteristics that are supposedly linked to T, it seems that existing beliefs about T are so strong and well elaborated that it might be difficult for both scientists themselves and their readers to see the machinations it sometimes takes to get the data to fit that belief.[12]

Recall that Kreuz and Rose abandoned their initial hypothesis when it didn't pan out. They then came up with a new hypothesis, using the only item in their data that did correlate with T: having been convicted as an adolescent for a crime they characterized as "aggressive or violent." This type of post hoc analysis is currently recognized as a major and widespread problem, especially in behavioral sciences; we will see that problem again later in this chapter, when we look at contemporary studies. It is by itself deeply problematic, but the clincher in this case is the way in which "violence" and "aggression" were defined. We don't get to see the raw data on criminal convictions, but we get the gist of them from table 4, which lists "adolescent convictions for more violent and aggressive criminal behavior in the 10 men with the highest and lowest levels of plasma testosterone." The table includes six separate offenses, five of which seem straightforwardly violent: first- and second-degree murder, attempted murder, assault, and armed robbery. A sixth offense, though, seems entirely out of place: "escape from institution." Their claim that men with the highest T values had more aggressive and violent convictions in their youth loses its force when you see that two of the top five T values come from men whose only "violent" offense in adolescence was to run away from a juvenile detention facility. Moreover, escape from juvenile detention accounts for one-third of the total "violent" offenses among these men with high T values.[13]

As it was reported in the *Washington Post,* however, the study took on a more ominous tone. Painting the men as having uniformly and extremely violent histories by naming only assault and attempted murder as examples, the paper reported that "'young violents' . . . secrete more testosterone than other prisoners." In addition to mischaracterizing their juvenile offenses, the wording seems to imply that high T was correlated in real time with committing violent crime. And what was the connection to the My Lai massacre and the conviction of Lt. William Calley? The *Post*'s reporter explained that "the Army researchers are trying to take these results and apply them, eventually, to the selection of soldiers." To wit, soldiers should be aggressive, but not too aggressive. In Rose's words, "A good soldier puts his energy to a task. He uses his aggression that way. . . . We don't want young violents. It's important for a solder to function in a group, not to go off and act aggressively on his own." At the height of protests against the war in Vietnam, this was more than a scientific statement; it was a political intervention that seemed to undermine activists' claims that it was an immoral war characterized by organized murder.[14]

The data from the Patuxent study are weak, but this specific enactment of T supports the story that the Department of Defense and the Nixon administration favored during Calley's trial: My Lai wasn't a gross failure of the armed forces, but the tragic and criminal result of an individual "young violent" who couldn't harness his aggression constructively. Using the term "young violents" to describe both the prisoners with higher T and out-of-control soldiers, Rose sweeps over the enormous gulf that exists between escaping from a juvenile detention center and the mass murder, rape, and mutilations committed at My Lai. He does so by linking them both with T, though neither Calley nor any of the other soldiers who committed the crimes at My Lai were actually known to have high T. It's not a scientific link—it's a narrative one.

Testosterone, Social Class, and Antisocial Behavior

"I doubt if there is enough testosterone in this entire audience to rob a single gas station," social psychologist James Dabbs once quipped in a talk at a social psychology conference. Dabbs, the eminent T expert from the

This American Life episode we describe in our introduction, was one of the most engaging writers and speakers ever to turn his attention to this hormone. Dabbs was a passionate researcher who sometimes reveled in debunking myths about T. Biographers have called him a champion of testosterone, which he characterized as "the most maligned, most misunderstood of human hormones." He was known, for instance, for his work suggesting that high T was associated not just with aggression and other negative behaviors but also with traits that go against the T stereotype, such as generosity and charm (especially when "good guy" behavior might increase social status). Nonetheless, when it came to the standard belief about T driving violence, Dabbs was strictly on the side of tradition. Consider his quip about his colleagues probably being testosterone-challenged. He knew little more about his fellow psychologists at the conference than their profession, but his assessment of their hormonal status telegraphed his beliefs about T, violent or antisocial behavior, and even social class.[15]

Dabbs picked up where Kreuz and Rose left off. Most of his research focused on T and aggression, including a large body of work on men and women in prison. In 1987 he published an article on the relationship between T and criminal violence among eighty-nine men in a Georgia prison. Prison staff had assigned men to their housing based on how "tough" they perceived the prisoner to be. To get a good mix of subjects, Dabbs recruited subjects from a "weak" dorm, a "tough" dorm, and the cellblock, which housed men considered too tough to control or too weak to defend themselves in individual cells. They then had guards and fellow prisoners rate each man's toughness. To get a thorough view of criminality, they looked at prison records, including types of convictions and parole board decisions, psychological test results, age, race, and more. To see if T was correlated with any of the measures of violence, they collected and measured salivary T. Spicing up their article with prison lingo to characterize the tougher men as "bo-hogs" and the weaker men as "scrubs," Dabbs and colleagues concluded that the "bo-hogs" had higher T.[16]

Even a cursory examination of the study's methods, though, reveals enough issues both to dispute its claims and to trace how the fact of T as a driver of violent crime is assembled. In Dabbs's day, there was hot debate about whether researchers should focus on violence or on aggression and about how these differ, as well as the best ways to measure either construct. Good scientific practice doesn't require that a research team resolve every foreseeable question before doing a study, but it does require a clear explanation that supports how particular measures fit with the abstract constructs they are investigating. But Dabbs and his colleagues instead presented their measurement decisions as if they were self-evident, using the subjective concept of observable "toughness" as the initial way to stratify subjects. Without information about how the term was defined for the raters—if it was defined at all—and how they applied it to prisoners, it is difficult to know what to make of this. Dabbs and colleagues don't explain whether a man's reputation for being "tough" in prison is supposed to be a measure of his capacity for or history of violence, his aggressiveness, or something else entirely. Similarly, they jumble together toughness, aggression, violence, and even the punishments that men receive as indicators of criminal violence.

To capture aggression, Dabbs combed through each subject's entire prison record, using multiple ratings and raters and taking into account each inmate's specific living context in prison. This enacts a view of aggression that is potentially all-encompassing: not just discrete criminal acts or extreme violence, but daily interactions with peers and guards, all focused on moments when the man had stepped outside the rules. Aggression is framed in this research as rule-breaking rather than violence per se. Dabbs uses an expansive concept that hinges on a man having a "deviant personality," rather than focusing on isolated behaviors.

One consequence of Dabbs's expansive view is that it generated an enormous number of potential links with T: the researchers made at least fifty-one different comparisons in this study, making it almost inevitable that some connections would turn up. Even so, only a meager four significant associations were found, three of which apply only to subgroups of the sample, and their explanations for why this is so are shaky. "Toughness" was associated with higher T only in the "weak" dorm. Dabbs reasoned that this was because the toughness ratings were most reliable in the weak dorm, but in fact the toughness ratings were extremely unreliable in all of the settings. Many men were rated as both "tough" and "weak" by different raters, and the reliability statistic was so poor that most researchers wouldn't find the measure good enough to use. Two of the findings that the researchers claimed as evidence of a link between T and violence applied only to men who were convicted of nonviolent crimes: among these men, higher T was associated with being punished more harshly for rule infractions and also with getting a longer sentence from the parole board. Dabbs and colleagues suggest that this might be because prison staff and parole boards know about men's "actual crimes," while the conviction on record might be for a nonviolent crime only due to plea bargaining. They imply that higher T might be a marker for "really" having committed a violent crime. If so, higher T would only differentiate men in the group that was officially listed as containing nonviolent offenders.

Dabbs found that "violent offenders" had higher T levels than "nonviolent offenders," which seems like a straightforward confirmation that T is related to violence. Yet there is another possible explanation that is just as simple, and based on well-established correlations: young men commit more violent crimes, and young men also have higher T. The researchers didn't control for age in their analysis, so the association between T levels and violent crime may very well be spurious. If so, being young is the thing that drives both higher T and higher crime. Higher T might be an innocent party here. Deciding not to control for age produces a particular causal chain out of the observations: age causes T to drop as a man moves from youth into middle age, and that lower T causes him to commit less violent crime. But controlling for age would allow you to see if there is any direct effect between T and violent crime instead of or in addition to the indirect effect through age. Dabbs's decision produces a version of the evidence that prevents consideration of that alternative interpretation. In the end, it looks like the Mulder effect again: Dabbs wants to believe.

Others want to believe, too. This is the single most frequently cited study on T and criminality in the literature, and it's still quite influential, considered a slam-dunk demonstration that T is linked to violent crime. In criminal justice textbooks, review articles, and original research in high-profile journals, researchers reach for Dabbs when they want to assert that the connection between T and criminal violence has been well established.[17]

The Criminal Class

By around 1990, the accumulated data on T and criminality were so fuzzy that even committed researchers like Dabbs conceded that the nature of the link between T and criminal behavior was "elusive." He suggested that T was at best a very weak predictor of criminality in part because no specific findings had been consistently replicated. But Dabbs wasn't ready to give up the hypothesis. Likewise, in his classic 1991 review "The Influence of Testosterone on Human Aggression," psychologist John Archer wrote that the existing human evidence for that relationship was "inconclusive" and presented evidence that the usual hypothesis might get the causal relationship backwards: instead of high T causing aggressive and competitive behavior, aggressive and competitive behavior seems to cause T to rise. Nonetheless, the review reads as if Archer found the evidence that T stimulates aggression among nonhuman animals to be so convincing that the same is likely to be true in humans, even if the effect is weaker, and moderated by more complex social variables. Everyone agreed, though, that evidence for an effect of T on aggression in humans was still sorely lacking.[18]

So, as is supposed to happen in science when evidence fails to confirm a hypothesis, the hypothesis pivoted. Up until that point, most researchers in behavioral endocrinology, at least those who were studying humans, followed a very simple model that led to the hypothesis that high circulating T predicts more masculine behavior. They expected that the more T someone had, the more "masculine" that person would be with regard to a particular trait, whether they were studying sexual behaviors, cognition, or aggression. But by the 1990s, behavioral endocrinologists studying other kinds of behaviors and personality traits had mostly moved into more complicated and nuanced territory—for example, considering not just circulating T levels in adulthood but T at critical periods of development, and understanding T as part of a suite of hormones that act together to influence and respond to behaviors.

Although violence researchers were slow to adopt these innovations, they began to abandon the idea of a simple correspondence between high T levels and violent crime. Those studying T and aggression didn't follow all the paths that other behavioral endocrinologists traveled, but they did make two key changes. First, researchers began to recognize that T doesn't just drive behavior but responds to it as well, as Archer had noted. We'll return to that development shortly. Second, researchers amped up their debates about how to model T's negative behavioral effects. Dabbs, among others, pointed out that "crime" is a uniquely human construct that is less about inherent attributes of behavior than about prevailing social norms and prohibitions. Reasoning that it might be expecting too much to see a clear and consistent association between "violent crime" and T, Dabbs and others shifted to a broader view of "antisocial" behaviors. Others, meanwhile, began to explore the idea that T drives not just aggression per se but a more general quest for dominance.

In 1990, a lengthy feature article on testosterone and aggression in men appeared in the *New York Times,* giving an unusually detailed picture of the scientific debate over how to measure aggression. In interviews, Robert Rose, James Dabbs, Allan Mazur, and other researchers jostled over the best way to capture the negative behaviors associated with high T. Most of these researchers believed that broadening the focus from aggression to dominance would yield the best payoff in T studies, because aggression is just one type of behavior that people use to assert dominance. Robert Rose, by then a psychiatrist at the University of Minnesota Medical School, said that T was most strongly linked to competitiveness and dominance behaviors, but he cautioned that the link between hormones and behavior in humans isn't direct. "In humans," he said, "hormones only set the stage, while social factors determine if and how they are expressed." Allan Mazur, who had been working on T for over a decade by then, explained that he favored a focus on dominance because "finding your place in the hierarchy is a basic part of primate life, and testosterone is tightly linked to the outcome of battles for dominance in other species." But dominance in humans doesn't look the same as in other species; instead, human dominance is usually "highly symbolic." Both Rose and Mazur had been studying how T levels can respond to changes in social status or in competition, and they were among the researchers contributing to the emerging view that social dominance isn't just a product of T but also drives T levels higher. Dabbs agreed that aggression was too narrow a concept to capture what T was doing, but he believed that dominance was the wrong way to broaden it. Instead, Dabbs thought T was related to "antisocial and aggressive behavior."[19]

One point of agreement was that social class must be considered to understand T's effects, but the researchers disagreed about how it fit in. Those researchers who had moved to the idea that "dominance" or "status-seeking" was the way to think about T believed that social class shapes the way men seek dominance, while T levels would affect the degree of that drive. Dabbs, on the other hand, indicated that T might actually be driving men's class position. He told Goleman that "many men with high testosterone levels are too impatient and aggressive to find their ways to positions of responsible leadership. Men with the highest testosterone levels were two and a half times more likely to be low-status as high." In other words, the other researchers thought high T drives dominance behaviors, but what dominance looks like would be different for men in different social classes. Dabbs, though, believed that high T may actually cause lower social class status, partly through criminality. These two views are very different, with Dabbs's angle looking like an update of Malthusian theory that social structure reflects the outcome of a struggle whereby the most "fit" rise to the top. But both versions of how social class fits into the relationship between T and violence deliver essentialist versions of the connection between violence and low social position, on one hand, and between "leadership abilities" and high social position, on the other.[20]

The story that social class shapes the form that dominance takes is assembled with help from the underlying idea that dominance behavior in high-status men is benign, productive, and socially valuable, while dominance behavior in low-status men is dangerous and often criminal. Describing T's role as supporting "the natural urge for the upper hand," the reporter suggested that the most dominant men in any situation are those with higher T, whether in "a prison yard or a boardroom." Paraphrasing the psychologist John McKinlay, the reporter wrote that "men high in the trait of social dominance tend to rise to positions of leadership in business and other organizations," while in other men "some traits associated with testosterone present an obstacle to success." Pause here to notice a temporal element: if traits are associated with T, and those traits lead to positions of leadership, then high T leads to being a leader—at least for some men. Men are either able to take positive advantage of high T or not. What is it that makes the difference? Social class. "For men of lower social and economic status, [high T] is likely to show up as a readiness for fights, a history of minor crimes, and chronic trouble with parents, teachers and peers in childhood. But that is not true at higher social and economic rungs, where the display of dominance is more subdued." Even without Dabbs's explicit view that T influences class position, the ability to express dominance in positive versus negative ways is a natural or inevitable feature of social class. There's a tension here between abstract theory and concrete research practices. In theory, T supports a homogeneous "masculinity," but in practice, T supports diverse and contradictory forms of masculinity, sometimes valued, sometimes dangerous, but all neatly aligned with existing social structures.[21]

The long-standing debate about the power of biology in determining behavior suggested a number of specific questions about T. Does T directly cause aggression and other antisocial behaviors? Or perhaps T fuels a universal drive for dominance, but dominance is expressed differently depending on social circumstances? The article is framed as a direct competition between these ideas. But there is something deeper. The question is not just what role biology plays in behavior but the role it has in social position. Does social position shape people's life chances and behaviors? Or, at least in a context where there are opportunities for social mobility, does a person's social position and status fundamentally reflect that individual's inherent qualities of personality and intellect? In other words, biology might not just shape behavior; it might also shape the composition of social classes.

To answer these questions, Dabbs and Morris used data that had been collected by the US Centers for Disease Control and Prevention (CDC) for a prior study of nearly 4,500 US veterans of the Vietnam War. The original study was a wide-ranging examination of the long-term effects of military experience for these veterans, but Dabbs and Morris scoured the dataset for all indications of what they called "antisocial" behavior, a broad category that encompassed everything from "having trouble with parents, teachers, and classmates" to assaulting other adults, going AWOL in the military, using drugs and alcohol, and having a larger number of sexual partners. Echoing the way that T and criminal violence were sutured together in the studies on prisoners, Dabbs and Morris use several strategies to assemble the connections between T, social class, and bad behavior. By stepping away from a narrow focus on criminality or even on violence to the broad notion of "antisocial behavior," thus exponentially expanding the number of variables and statistical analyses, they almost guaranteed that they would find some relationship. They concluded that high T is a direct cause of antisocial behavior, but that the relationship is strongest among men of lower socioeconomic class.[22]

Dabbs and Morris's statistical analysis left a lot to be desired. Instead of regression analysis, they compared the behaviors of "high T" versus "normal" groups. To see how social class fit in, they split the men into high and low social class groups, using just education and current household income—a measure that inherently obscures the role of early influences like parents' social class. Further, instead of checking for a statistical interaction with class, as would have been standard, they simply repeated the analysis of behavior and T in each of the class groups. Perhaps the biggest problem, though, is that Dabbs and Morris didn't acknowledge the class bias inherent in military service: the burden of military service during the Vietnam War fell harder on young men who were poor and working-class. Among service members, lower-class men were more likely to see combat. The complex, negative sequelae of combat experience might explain the associations they saw among class, negative behaviors, and T. Still, Dabbs and Morris expressed uncertainty about whether T's effect was "cognitive" or "motivational," but they had no reservations about their conclusion that T is especially problematic among lower-class men.[23]

By the time Dabbs's popular book *Heroes, Rogues, and Lovers: Testosterone and Behavior,* was published, his views about high T levels driving social class position had crystallized. Testosterone gives a modern, scientific explanation for the perennial trope of poor and working-class people as prone to deviant behavior. The book is seasoned throughout with colorful working-class characters—he turns repeatedly to construction workers, plumbers, steelworkers, and other blue-collar folks who exemplify the "rambunctious and impatient" tendencies of high-T people. While the trope is threaded throughout the book, it isn't woven into a solid or consistent argument that T is firmly linked with lower class position. On one hand, he explains the fact that T is more strongly related to antisocial behavior among working-class men by suggesting that higher-status people have more to lose from following their "impulses." Likewise, he devotes considerable time to high-status jobs that he says are linked with high T levels, especially the job of trial lawyer. On the other hand, he suggests that it is high T itself that lands people in low-status occupations, writing that the traits of high-T people are generally more "useful to those in rough-and-tumble occupations, occupations that are often associated with low social status." Moreover, the "tendencies" of people with high T levels "interfere with getting an education, which is usually essential to white-collar success. High-testosterone boys don't like to sit and listen to the teacher day after day, and high-testosterone men find most white-collar work boring and confining. This makes it difficult for them to hold on to high-status jobs. Their excessive competitiveness might also interfere with white-collar success, where being a 'team player' is at a premium." The elitism in the literature on T and behavior is especially pungent in Dabbs's work.[24]

There's a narrative strategy in the book that we call "pastiche science." Particular studies are interwoven with a range of individuals' stories, including living people, historical figures like Jesse Owens and Joan of Arc, and even fictional characters like Sam Spade. Dabbs uses the characters to add flesh and bone to the skeletal connections between T and personality traits or behaviors, but there is a problem: in the vast majority of cases, there is no information on the T levels of the people he describes, and it is often absurd to think there would be. There is only speculation and association. This method of placing stories side by side in order to create meaningful links is similar to what the *Washington Post* did by juxtaposing William Calley's trial report and Kreuz and Rose's research. Presenting the stories together hints at their interchangeability and implies a stronger causal chain than is suggested by any single study or story: it isn't necessary to have all the variables for any single story. Granted, *Heroes, Rogues, and Lovers* was written for a general audience, but this strategy is emblematic of what M'charek and colleagues describe when they write that facts are crafted out of "heterogeneous elements." Here, the "facts" are seemingly about T, but the elements inextricably embed meaning about social class. In particular, class as it is crafted in Dabbs's work fits the American mythology of class mobility: your temperament and behavior more or less land you with the status you earn and deserve.[25]

Honor Cultures: To Be Young, Violent, and Black

While Dabbs's work uses T to cement folk stories about violence and disorder among working class people, Allan Mazur achieves a similar effect by using stories about race and T to give scientific weight to the trope of innate black criminality. Mazur was one of the first researchers to document that social contexts affect T levels in humans. Formally, his work seems to eschew biological determinism for a nuanced model that includes reciprocal interactions between environment and bodies. Unlike behavioral genetics, in which correlations between behavior and race conjure an obviously essentialized version of race, Mazur's studies seem to be premised on an idea about race as a social environment. Thus, when he finds that T levels correlate with racial categories, it looks compatible with the observation of social epidemiologists and others who describe how race becomes biology: "We literally embody the world in which we live, thereby producing population patterns of health, disease, disability, and death." But these articles are deeply unsettling in the way they uphold classical arguments for white supremacy, purportedly giving credence to the notion that black people are inherently disordered and dangerous.[26]

Mazur's work is predicated on assumptions about T and behavior that took hold at the turn of the twenty-first century. Researchers had been puzzled and frustrated for decades with findings suggesting a weak and inconsistent relationship between T and aggression in humans. Rather than pursuing different explanations for aggression, or alternative roles for T in human bodies and lives, many researchers succumbed to the Mulder effect: they wanted to believe that T caused people to be aggressive.

Newer models connecting T to human aggression borrow from research on birds. By the late 1980s, researchers had amassed a huge database of information on T secretion in the males of more than twenty bird species. The extent and kinds of variation in T were intriguing: seasonal shifts in individual birds' T levels, discrepancies between levels and patterns in free-living versus captive populations, diversity in the way T correlates with behavior. One of the biggest puzzles, from the researchers' perspectives, was the inconsistent evidence about the relationship between T and aggression. In a landmark 1990 paper, the zoologist John Wingfield and colleagues concluded that differences in social context might explain the discrepancies, and proposed that both T and aggression rise together in response to challenges from other birds. They called this proposition "the challenge hypothesis," explaining that the relationship between T and aggression is strongest when there is "social instability, such as during the formation of dominance relationships, the establishment of territorial boundaries, or challenges by a conspecific male for a territory or access to mates." But the levels of aggression, and the link between T and aggression, both decline "during socially stable periods and when territories have been established, with status or boundaries maintained by social inertia." The process reflects seasonal variation, too, in that more aggression is observed during the unstable mating season, but once the hatchlings are born, aggressive behavior settles down. Finally, it's a mutually reinforcing loop. High T apparently primes males to respond to instability and conflict with more aggression, which in turn keeps T high. But once the social relationships stabilize, with settled territories and pairs, T drops, which also helps calm the atmosphere and reduce future challenges, so T stays low until the next breeding cycle or some external event changes the equilibrium.[27]

In a sweeping 1998 review, published with more than two dozen invited commentaries, Allan Mazur and his fellow sociologist Alan Booth implicitly drew on the new challenge hypothesis as they laid out the evidence on testosterone and dominance in men. Their goal was to synthesize the emerging theories that T and behaviors affect each other reciprocally and that, in humans, researchers should not focus narrowly on frank aggression but look more broadly at dominance. The article is now twenty years old, but it's the most frequently cited article in the scientific literature on dominance, aggression, or violence and T among humans, and it's still the go-to think piece for many researchers.[28]

Mazur and Booth suggested that the evidence to date was weak because researchers had been using the wrong model. Kreuz and Rose, Dabbs, and others researching aggression in humans had been using what Archer later called the "mouse model," which predicts a "simple causal relationship between circulating testosterone and aggression." Dabbs's attempts to incorporate social environment into the model look primitive compared with this new scheme. The key difference is that Mazur and Booth incorporated the idea of reciprocity, emphasizing that baseline T levels affect behavior, but that the social environment and behaviors also affect T levels, so there is an ongoing mutual influence. They especially drew on Mazur's studies of competition, arguing that T rises in anticipation of competition, and those who win a competition get an additional burst of T. They speculated that some social environments are like an ongoing dominance contest, characterized by constant social challenges and instability.[29]

They called these challenging and unstable environments "honor cultures," a concept that sutured challenge-hypothesis research to racial discourses on violent "subcultures." Mazur and Booth borrowed the concept of "honor cultures" from the psychologists Richard Nisbett and Dov Cohen, who coined the term to argue that in the United States, a southern "culture of honor" could explain why "argument-related homicide" rates are substantially higher for white southern men than for white northern men. T was implicated in this phenomenon when researchers showed that T rises when men from the South encounter even minor insults, in contrast to no rise in T seen among northern men. Mazur and Booth summarized that research as showing that men from the South are hypervigilant about insult, often reacting dominantly and even violently to perceived slights. But they quickly moved from a theory that was about white southerners to a more general hypothesis about the relationship between features of subcultures that are, or were historically, dominated by young men: "There may be a general hypersensitivity to insult in any subculture that is (or once was) organized around young men who are unconstrained by traditional community agents of social control, as often occurs in frontier communities, gangs, among vagabonds or bohemians, and after breakdowns in the social fabric following wars or natural disasters. When young men place special emphasis on protecting their reputations, and they are not restrained from doing so, dominance contests become ubiquitous, the hallmark of male-to-male interaction."[30]

In spite of that generalizing move, they didn't explore honor cultures generally, or even a range of so-called honor cultures. Instead, they lifted the concept to build a biosocial argument about violence among young black men. Mazur and Booth argued that the concept of honor cultures precisely describes the "subculture" of "poor young black men," building their case via an extended quote from "The Code of the Streets," an essay by the African American sociologist Elijah Anderson that appeared in the *Atlantic* in 1994:

> Most youths have . . . internalized the code of the streets . . . , which chiefly [has] to do with interpersonal communication . . . , [including] facial expressions, gait, and verbal expressions—all of which are geared mainly to deterring aggression. . . .
>
> Even so, there are no guarantees against challenges, because there are always people looking for a fight to increase their share of respect—or "juice," as it is sometimes called on the street. Moreover, if a person is assaulted, it is important, not only in the eyes of his opponent but also in the eyes of his "running buddies," for him to avenge himself. Otherwise he risks being "tried" (challenged) or "moved on" by any number of others. To maintain his honor he must show he is not someone to be "messed with" or "dissed." . . .
>
> The craving for respect that results gives people thin skins. Shows of deference by others can be highly soothing, contributing to a sense of security, comfort, self-confidence, and self-respect. . . . Hence one must be ever vigilant against the transgressions of others or even appearing as if transgressions will be tolerated. Among young people, whose sense of self-esteem is particularly vulnerable, there is an especially heightened concern with being disrespected. Many inner-city young men in particular crave respect to such a degree that they will risk their lives to attain and maintain it.[31]

Mazur and Booth suggested that the reciprocal model of T, together with certain features of social life in urban black communities, could explain the problem of violence in the "inner cities"—a problem they figured narrowly as struggles among young black men. Their story goes like this. Perceiving minor insults as serious challenges, young black men's bodies react with a spike in T. This higher T then makes it harder for them to walk away from challenges; the resulting fights cause their T to stay high, trapping them in a "vicious cycle" of violence driven by T. Mazur and Booth claim that honor culture is the most important dynamic affecting

T for these young men. Finally, they argue that their hypothesis is supported by evidence that black men have higher T levels than white men, based on the same data from US Vietnam-era military veterans that we described earlier.[32]

Mazur and Booth's argument swiftly rehearses pathologizing narratives about black communities and young black men in particular that are so familiar it can be easy to miss the ways these connections get made. They carefully set up a narrative that allowed them to disavow racism, even as it was built in through associations. Throughout the passage, Mazur and Booth used the terms "the street," "inner cities," and "poor young black men" interchangeably, and repeatedly used "the street" as both a setting and a substitute for "young black men." They relied on the racial coding of "the street" and "inner cities" as black, and on the dominant racial imaginary that figures black communities as characterized by "breakdowns in the social fabric" and as lacking "traditional community agents of social control." They didn't have to spell it out, because the white racial imaginary does that for them. They opportunistically used Elijah Anderson's words to paint a picture of individuals who are pathologically sensitive to insult: thin-skinned, craving respect yet insecure, on a hair trigger for a fight, and irrationally attached to reputation. Because Mazur and Booth didn't refer to any other aspect of the social context, the only "challenges" that young black men face seem to be coming from other young black men. Here and in later works, Mazur and Booth used this vivid ethnographic passage—complete with "street jargon"—but consistently excluded Anderson's larger framework, which includes "a legacy of institutionalized racism, joblessness, and alienation."[33]

Though the discussion has the feel of an empirical study, it's nothing more than a pastiche. Mazur and Booth presented no data on violence or crime among black men, nor on the nature of the stresses young black men face. They presented no data on T levels for young black men who are supposedly a part of these honor cultures. Most of the elements of the analysis are simple speculations, stitched together by familiar stories about race and by a science-y style of writing that seems to take alternative explanations into account but actually does not. Other than a single mention of poverty, there is no mention of structural factors like chronic unemployment. In an argument that is supposed to be about a challenging social environment, Mazur and Booth fail to consider the general or specific ways that young black men are systematically devalued and thwarted in educational, economic, and social terms under white supremacy. All the challenges are self-imposed.

The glue that secures Anderson's material to their biosocial argument is the claim that black men have higher T than white men, based on data from the same study of Vietnam-era veterans that Dabbs and Morris used to make claims about T, aggression, and social class. Mazur and Booth cited a racial difference in veterans' T levels, saying it confirmed their honor culture hypothesis: "Only among *younger veterans with little education* do we find T in blacks to be unusually high, significantly higher than in whites. These younger black men, poorly educated, most of them urban residents, are most likely to participate in the honor subculture, and that may be the reason for their elevated T." But just as they didn't have data on the actual behavior or T levels of young black men in urban settings, they also didn't have the requisite data to support the story they told about racial differences in T among these veterans. Most importantly, there were no data about where men in that study reside. Mazur and Booth used current income and education to infer where the men lived, assuming that black men with more money and education would move out of cities. They further assumed that young black men in urban settings would be "most likely to participate in the honor subculture." In Mazur and Booth's telling, which relies more on a sequence of inferences than on data, "honor culture" becomes a phenomenon unique to black men. References to T make it seem like more than familiar racial generalization.[34]

It's interesting to consider why the blatant racialization in this article and in Mazur's other pieces that repeat the analysis has never been called out. It might be because the biosocial model provides cover, explaining racial differences in T by reference to an interaction between individuals and their environments rather than an essential difference in racial biologies. But the biosocial model has a suspiciously neat fit with dominant racial discourse. For instance, they argue that "antisocial actions are often attempts to dominate figures in authority (teachers, policemen) or, more abstractly, to prevail over a constraining environment," giving examples like schools, the military, prisons, and even families. Conceivably, you could drop anybody into these environments and you might see the same relationships irrespective of characteristics like gender or race. Yet their extended examples always involve black men.[35]

The overall result is a bait and switch: it reads as though they're talking about environments and social context, but their discussions are grounded on an essentialized notion of racial difference. Honor cultures provide a framework that figures black communities as sites of utter chaos, and black men in particular as uncivilized.

This foundational article was coauthored, but a deep dive into the racialization narrative points to the influence of Mazur over that of Booth. Mazur had already planted the seeds for the black "honor culture" hypothesis in 1995 in an article exploring hormones and "deviance" among US army veterans in the massive CDC dataset used in multiple studies mentioned previously. In that article, Mazur explained that deviance, defined as violating conventional norms, is an important form of human aggression that "is especially likely in the absence of effective social controls and among individuals who are not well integrated into society's mainstream." The reference to "absence of effective social controls" is not explicitly racialized in this article but elsewhere is directly linked to black communities, especially by pathologizing female-headed households. Mazur gives special attention to one of the dozen or so analyses he conducted on the data, one that examines the simultaneous effect of education, age, and race on T levels and showed that young black men with low levels of education had the highest T levels among all the groups. Having also found that black men had higher scores on several measures of deviance, Mazur suggests that "possibly the extremely high testosterone levels of young, poorly educated black males contribute to deviance in that cohort." Yet he has no data showing that young black men are more "deviant" than older black men, who have the lowest T among the four groups. That "possibly" is important because Mazur has a strong agnostic tone in this article regarding the direction of causation between hormones and deviant behavior: citing his own and Booth's work on sports and other competitions, he is firm that the influences between hormones, on one hand, and behaviors and situations, on the other, flow in both direction. Yet based on multiple statistical models, he ultimately concludes that the results are consistent with the hypothesis that high T is driving "deviant actions," rather than the other way around. He lands on this conclusion despite the fact that midway through the article he presents statistical results showing that education and income are the factors that seem to explain deviant behavior best, while "hormones do not explain much of the deviant behavior of young black males."[36]

Although Mazur is known for his commitment to and empirical support of biosocial models, he has for over two decades been crafting an essentialist narrative of aggression among young black men, especially those with low education levels. Whereas racialization is explicit in Mazur's work, most studies of human aggression or violence don't highlight race at all. That doesn't mean race isn't operating in these studies, but piecing together how racial associations emerge requires looking within individual studies and also across them. For example, by focusing on aggression and violence only when samples are predominantly or exclusively lower-class and/or people of color, while the nonaggressive forms of dominance that researchers consider to be more common in humans are explored in predominantly white, highly educated samples, researchers may inadvertently activate racial tropes of aggressive black men and boys.[37]

Meanwhile, the racialization in Mazur's work has become even more explicit over time. In a 2016 paper, Mazur revisited his hypothesis from the 1995 paper on veterans and again reported that black men in their twenties with a high school degree or less have "inordinately high T." Mazur takes two big leaps to use these data as evidence of black men's ostensible propensity to aggression. First, he again frames the whole analysis around "inner cities" and "honor cultures," even though he has no data on where the men reside, and in fact cites evidence that only about half of US black men with a high school degree or less live in cities. Second, with no information on the behavior of men in this study, he nonetheless suggests that they are prone to physical aggression, again leaning on Elijah Anderson's ethnography from more than two decades earlier. The broad links he draws between "inner-city" violence and high T among black men are all the more problematic given that he routinely asserts there is no direct link between T and aggression or violence.[38]

Zombie Facts

For as long as there has been research on T and human behavior, scientists have been trying to confirm that T drives human aggression. Yet the best evidence is still solidly rooted in 1991, when both James Dabbs and John Archer deemed it "inconclusive." How, then, can this dead fact continue to live on?

The zombie fact that T causes human aggression is animated and perpetuated by a number of strategies, such as the move to more sophisticated and complex research models. Newer models such as the challenge hypothesis better align human research with studies in other species, but these models also introduce more interpretive flexibility. Rather than clearly clustering with one of the newer models, positive findings continue to be scattered across specific hypotheses. But shifting across models allows investigators to revisit data time and again to search for the expected connections to T. For example, researchers are still struggling to fit cortisol into their models, with some suggesting that T is linked to aggression only when cortisol is low, others rejecting this model, and still others suggesting that additional chemicals like serotonin or factors like prior experience with violence have to be taken into account for the influence of T to emerge. We haven't even lingered on the effect of these shifting enactments of T. Is the T that matters for aggression one that emerges only in the company of other hormones or neurotransmitters like serotonin? One that acts most decisively in a critical prenatal period? One that's active in adulthood and therefore best studied through circulating T? The questions go on and on. Tracing the multiple, non-aligning enactments of T through these studies would reveal even more distance between the hypothesis that T causes aggression and the evidence for the claim. Shifting Ts, methodological maneuvers, and conflicting models fly under the radar because the findings are so irresistible.[39]

The zombie fact that T causes human aggression, especially criminal aggression, is also kept alive by compelling storytelling. Research on T and criminality or aggression persistently points to broad, important social problems as the ultimate reason for doing the work, and asserts the importance of the knowledge that it will generate toward solving these problems. The studies and supporting literature use predictable narrative structures and terms to signal authenticity and authority about worlds that are separate from the experiences of their target readers, assumed to be elite and white: think of the "bo-hogs and scrubs" in Dabbs's prison research, or Mazur's repeated use of Anderson's description of young black men seeking "juice." In both cases, the storytelling traces relationships where the data don't coalesce, creating a pastiche that reads like a set of causal connections. Just as Mazur projects urban status and "honor culture" membership onto young black men with high T levels, Dabbs's stories project high T onto anyone—real, historical, or fictional—who manifests any attribute suggestive of T. Some of these characters don't even exist, and even for the real people who are referenced, there's often no evidence of their T levels, but these fictions never register.

Newer stories in T research gain a good deal of their power from habitual citation of classic studies such as those described in this chapter, which to date have escaped criticism. In those studies, loose statistical practices that would now be identified as "*p*-hacking" or fishing, shifting hypotheses in the middle of data analysis, using measures with poor reliability (like the toughness ratings in Dabbs's prison studies) and poor validity (such as counting escape from juvenile detention as a violent crime), and dozens of analyses on small samples made it inevitable that the researchers would turn up connections between T and violent aggression. The claim that T causes human aggression travels outward from strict aggression-focused work and radiates to every other domain of T research. Distanced from the particular, often dubious contexts in which it was produced, the assertion that "T increases aggression" can anchor other hypotheses, such as those about high T increasing risk-taking or, conversely, low T being beneficial for new fathers. We look closely at both those ideas in later chapters.

Ongoing circulation of the idea that T underwrites human aggression isn't a trivial matter. It contributes to a hyperindividualized understanding of aggression and crime at the same time that it amplifies essentialist frames that hitch aggression to specific human groups: males in general, working-class people, and people of color. The way these connections are made has changed over time. The Kreuz and Rose study and the way it was positioned in the public discourse made the focal shift from social problems of violence to individual bodies look simple and straightforward. It advanced a narrative that violence, even in the context of war crimes, is not a reflection of anything inherent to a culture or its institutions, not the military or even war itself. The problem is young violents, specifically young men who can't control themselves because they have too much T.

Fast-forward fifty years, and it's much harder to see how scientific research on T and violence works in our cultural narratives. It looks instead like there is a big rift now between the pop-culture idea that T causes violence and seemingly more complex and nuanced scientific ideas about T. This is especially true because of the advent of biosocial models. Yet biosocial researchers studying T and dominance behaviors tend to consider only the micro-contexts of face-to-face interactions, as if these were sealed off from broader social processes and institutions. None of the T studies we've seen consider how the racial and class composition of their samples are related to social institutions and the operations of power, and how these in turn shape the lives of the people they study, whether they are looking at prisoners, military veterans, schoolchildren, or business school students. Instead, the studies follow folk notions and reiterate the social canalization of power.

In twenty-first-century research, T is still available to deflect a structural, institutional analysis in favor of individual biology, but this time studies aren't underwritten by a simple model of a troubled lone wolf who "hasn't learned how to control his own aggression." Instead, current studies deploy more complex models that put T in the mix with multiple chemicals, and they call on models positing that aggression developed in early humans as an "adaptive" response to the challenges of their environment but over time has become non-adaptive in modern humans. With this narrative come various ideas about how human behavior can "relapse," such that "more ancient aggressive behavioral predispositions become adaptive once more," as social neuroscientist Jack van Honk and colleagues put it when explaining how aggression in war is different from aggression in "peaceful" conditions. Their model looks especially up-to-date because it centers on "the emotional brain" and casts T, cortisol, and the neurotransmitter serotonin as the "core brain chemicals of reactive aggression." But look more closely at the construction of this neuro-friendly model and you will find a familiar rock pile of old James Dabbs studies. Citing only the prison data that we examined earlier, van Honk and colleagues sidestep the problem that most research has found little to no link between T and human aggression: "Testosterone levels have repeatedly been associated with observer-rated conduct disorder problems and violent reactive aggression in large populations of both males and females." Indeed, "the studies of James Dabbs . . . leave little doubt about testosterone being implicated in reactive aggression in humans." They also don't seem troubled by the fact that Dabbs wasn't looking at the specific version of T that van Honk and colleagues insist upon, which is high T in the context of low cortisol—more evidence that T's multiplicities are sometimes exploited to wring more significance out of data and sometimes buried when it is more expedient to look at T in isolation.[40]

The sociologists Steven Barkan and Michael Rocque have recently pointed out that biological and biosocial models now dominate criminology. Citing the "troubling" fact that "the last few decades have seen a profusion of individual-level theories of crime," Barkan and Rocque demonstrate that criminologists' preoccupation with "proximate causes of antisocial behavior" is coupled with turning a blind eye to factors like poverty and racism. T plays an important role in the field's eager embrace of biosocial and cultural factors. As an example of the biosocial trend, Barkan and Rocque point to a late 2018 review of the "race, poverty, and crime nexus" that aims to "spoil the politically correct mantra that black crime results from white racism." In that piece, which appeared in the top-tier *Journal of Criminal Justice,* sociologists Anthony Walsh and Ilhong Yun devote a substantial chunk of their text to van Honk and colleagues' "triple imbalance theory," which they group with Mazur and Booth's "honor culture" hypothesis about T and criminal aggression among young black men and with psychologists Lee Ellis and Helmuth Nyborg's report of higher T levels among black men compared to white men. With narrative strategies that recall the opportunistic use of Elijah Anderson's work by Mazur and Booth, Walsh and Yun dismiss structural explanations for crime by adding Asian Americans to the mix with a simplistic analysis that assumes racism is the same for all nonwhite groups, though it is well established that different racial and ethnic minority groups are differently racialized. In the end, their argument is this: Asians have also experienced discrimination and racism; anti-black racism in America "was" Jim Crow, but that has been "over" for more than fifty years, while "anti-white" racism by blacks is now more serious and pervasive; Asians have lower rates of criminality than whites, while blacks have higher rates. Therefore, if racism explains crime, "proponents would have to attribute Asian successes and low crime rates to pro-Asian bias on the part of whites to their own detriment." Among the many astonishing elements of this paper is that a reputable sociology journal would publish such a narrow and distorted analysis of racism, which rests on a selective and legalistic reading of racial statistics ranging from political representation to out-of-wedlock births to crime statistics and much more, but omits all evidence that structures of racism go far beyond legal discrimination to encompass, at a minimum, enduring practices and economies that privilege white people. It is especially egregious that an article on race and crime in 2018 would fail to engage the evidence of decades of intensive policing in black communities and the dramatically harsher treatment that black people receive in all stages of the criminal justice process. Time will tell if this particular article gets any traction, but it's important to recognize the political and rhetorical importance that a long legacy of studies on T and criminality have in the scientization of current white supremacist views.[41]

While specific hypotheses shift over time and research populations are adjusted, T's powerful connection to masculinity has persisted. What falls under the category of "masculine," though, has been more mutable. The studies of aggression make it clear that T facilitates forms of masculinity considered dangerous and devalued as well as healthy, valorized ones. T is flexible enough to accommodate multiple masculinities. This flexibility was formalized when researchers began to theorize that T fuels dominance, while dominance takes different forms in specific contexts. From that point forward, race and class were embedded as "ghost variables" in this literature, usually not present on the surface but creating potent meanings through the use of culturally loaded terms, sampling strategies that reinforce class and racial stereotypes, and inferential leaps that go unnoticed because they fit cultural expectations. In this pastiche science, the variables in the causal chain are rarely measured in tandem. Instead, the link between T and violent aggression is made through juxtaposition of multiple stories that include partial associations: over here, high T is linked to young black men; over there, young black men form an "honor culture"; at the population level, violent crime is connected to young black men. Only by weaving together a web of material and symbolic associations does this become a singular story about T, violence, and race. Likewise, the plumbers and football fans and construction workers who populate James Dabbs's work put warm flesh on the cold assertion that working-class people do the jobs they are suited to, but there's nothing in his narrative that fills in the larger cultural, economic, and social forces that have directed them into those jobs. As Amade M'charek has described in thinking about genetics and race, fact and fiction are made of the same "stuff." In this literature, we see all three functions of race that M'charek has observed: it is simultaneously an object of study, a method for categorizing, and a theory of differences and how they come about.[42]

•   •   •

CRITICAL ANALYSIS OF HORMONE research has mostly treated hormones as a technology for making gender, but here we have shown that hormones also make race and class. The resurgence of race as biology is typically attributed to the rise of genomics and the routine slippage between racial categories and other ways of grouping genetic information, such as by haplotype or population. Research on hormones also facilitates this resurgence. The endocrine mode of racialization may be even more pernicious because it is harder to trace, requiring that you go beneath the surface of studies. The current dominant model in behavioral endocrinology of aggression, the challenge hypothesis, plays a key role in obscuring how race is essentialized in these studies. Researchers seem to begin with the race-neutral idea that humans, like other animals, respond to social challenges with a spike in T, and that this spike makes aggressive behavior more likely. In the abstract, this model presents the exciting possibility of empirically demonstrating the intimate entanglement of bodies and their social worlds, potentially connecting threads that are as disparate as behavioral endocrinology, social epidemiology, and feminist science and technology studies. But in practice, the process of embodiment—that moment where the social is imbibed and transformed into the biological—is not under investigation. In its place is a thin, shallow conceptualization of the social, which figures in the research as an assumed and homogeneous background to subjects categorized by race or class: race isn't connected to social institutions and history, but is a collection of habits. This biosocial theory enacts blackness as membership in a disorderly "subculture" that generates specific challenges for its members. Blackness is the source of the challenges in this challenge hypothesis.

The way that hormones make social class might be the biggest surprise coming out of this analysis, and it could well be the most important lesson. We do not have adequate tools for thinking about social class and how class divisions are produced, justified, and maintained in contemporary science. Close analysis of the way that studies in behavior biology, whether based in endocrine, genetic, or other models, build in class-based elitism is long overdue. Studies of T and aggression suggest that T might interact with social position to make some people unable to assume the responsibilities of biological citizenship. This is new packaging but leaves working-class and poor people nearly as biologically unfit for modern civilization as did nineteenth-century theories of atavism among the working classes. At this historical moment, when the importance and volatility of class relations play such a powerful role in politics, it is especially urgent to shine some light on the casual contempt for poor and working-class people in the making of facts about T and violence.

Research linking T to convictions for criminal violence, prison guards' assessment and treatment of prisoners, and parole board decisions makes crime and punishment seem like an objective phenomenon that is mostly driven by the characteristics and behavior of individual people who are arrested and incarcerated. It is not. In the United States, arrest and incarceration are irrefutably bound up with social class and especially with race. Racial disparities in criminal justice policies and practices are arguably the single most powerful institutional factor that disadvantages people of color in America today.[43]

It might not seem so important to take these arguments on board in thinking about contemporary T research, when no high-profile researchers are actively investigating the link between T and criminal violence. In *Ghost Stories for Darwin*, botanist Banu Subramaniam has elegantly demonstrated how the theory of evolutionary biology and research practices in the field incorporated eugenics from the start. Revisiting her own early studies, she considers how that residue directed and constrained her research, and in so doing, she models the work necessary to take research in radically new directions that don't inadvertently carry forward and even amplify that legacy. Subramaniam's example suggests it's necessary and possible to do the same sort of excavation for research on T and violence.[44]

4

POWER

THE SECOND-MOST-WATCHED TED talk is by Amy Cuddy, a social psychologist at Harvard Business School at the time, on what she calls "power posing." The talk, "Your Body Language May Shape Who You Are," had over 52 million views as of this writing. Engaging and upbeat, Cuddy tells the audience she's offering them a "free, no-tech life hack, and all it requires of you is this: that you change your posture for two minutes." Just strike a few powerful poses and, well, you'll feel powerful, she beams, explaining how she and her colleagues Dana Carney and Andy Yap uncovered this surprising effect in their lab. Cuddy widely disseminated their findings, generating international attention and putting her in high demand for media appearances, speaking engagements, and corporate training.[1]

Social scientists have long known that we make sweeping judgments about each other based on body language, or what Cuddy calls "nonverbals." "And those judgments," she notes, "can predict really meaningful life outcomes like who we hire or promote." But our body language does not just influence others, she says: nonverbals shape our own thoughts, feelings, and physiology. Research on power and dominance, whether animal or human, she goes on, shows us that those in power use "expansive poses"—big, limb-stretching, chest-raising gestures that take up space. She illustrates the point with a series of photos: a grumpy-looking gorilla, a swinging orangutan kicking up one foot, a hooded cobra, a swan with wings spread, Oprah at a desk with her hands clasped behind her head,

Mick Jagger with arms raised. The final photo is Usain Bolt, his long arms stretched up to form a V at a race's finish line.

Powerful people, Cuddy tells the crowd, are more confident and assertive and take more risks than powerless people. In addition—and the reason we are interested in this research—they also have higher testosterone, which she declares to be "the dominance hormone," and lower cortisol, which she calls "the stress hormone." In the animal kingdom, Cuddy reports, high-power alpha male primates have high T and low cortisol. What's more, if there's a power vacuum at the top of the chimp hierarchy and a lower-echelon male takes over, the new top chimp will exhibit significant increases in T and decreases in cortisol. What does this mean? asks Cuddy. Could people fake a role change and feel more powerful as a result? Could how we hold ourselves change our thoughts and feelings, the physiological factors like hormones that shape them, and ultimately how we act? Cuddy and colleagues conducted a study and concluded that yes, we can. Drawing the audience in with her research question, she directs them to employ a "tiny intervention": "For two minutes . . . I want you to stand like this, and it's going to make you feel more powerful." Cuddy holds the familiar Wonder Woman power pose—chest out, hands on hips, feet planted firmly in a wide stance. Maintain this position for two minutes and your T will go up, making you feel powerful, she explains; your cortisol will go down, making you better able to tolerate risk. Power posing, according to her research, "govern[s] how we think and feel about ourselves," helping a person to summon the confidence that's advantageous, for instance, in an important job interview or a stressful presentation. One can't help but conjure an image of bathroom stalls everywhere cloistering a secret army of corporate managers before their annual reviews, hands on hips, chests inflated, hoping for the best.

In the emotional high point of her talk, Cuddy shares the personal inspiration behind her research. At nineteen, Cuddy was critically injured in a car crash. Post-accident, people told her she wouldn't be able to finish college because of the cognitive impact of the injury. It was a huge blow to her self-identity as a smart person, and she felt "entirely powerless." She did complete college, though, and went on to graduate school, where she found herself feeling like an imposter. When she expressed a strong urge to quit, her mentor cheered her on. "Fake it till you make it," she told Cuddy, who followed her advice, finished the program, continued her studies, and landed a job as a professor at Harvard Business School. There she saw her own earlier insecurities mirrored in her women students, who, as a group, participated much less than men, even though they made up half of the classes. She recalls a particular student who hadn't said a word all semester. Cuddy told her she had to participate or she'd fail, and the woman confessed, "I'm not supposed to be here." This was Cuddy's aha moment. Though she no longer felt that way herself, she recognized this particular shame and beseeched the student, voice rising, "You *are* supposed to be here! And tomorrow you're going to fake it, you're going to make yourself powerful." The audience breaks into vigorous applause.

Cuddy's advice worked. "She comes back to me months later," she wraps up, "and I realized that she had not just faked it till she made it, she had actually faked it till she became it." She directs her audience in the now familiar pep talk, adding: "Do it enough until you actually become it and internalize it." The takeaway is emblazoned on the screen: *tiny tweaks* → *BIG CHANGES*. Her final mandate to the audience is blunt and specific: "Get your testosterone up. Get your cortisol low." Strike a pose, rule the world. Or at least your world.

Cuddy went on to parlay her TED talk into a bestselling book, and later into the central tool in her motivational consulting, literally selling "power posing" as a social justice technique backed by science. She urged anyone feeling a paucity of "power" in nearly any situation to use the method: in a classroom or a job interview, but also before getting on the dance floor. "The people who can use it the most," she says, "are the ones with no resources and no technology and no status and no power . . . and it can significantly change the outcomes of their life."[2]

Cuddy has amassed testimonials from people who assert that power posing makes them feel more powerful. We aren't questioning people's feelings or the value they see in power posing. The testimonials, however, viewed alongside the study and its critics, raise an intriguing question: How much does the power in power posing hinge on "getting your testosterone high" and getting "your cortisol low"? Our conclusion may not come as a surprise: It doesn't. The whole story isn't that simple, though, and it's worth carefully tracing the "Now you see it, now you don't!" maneuverings of T in this whole operation, including how the hormone figures (or doesn't) in the withering criticisms from other scientists that would come to plague, and ultimately topple, this research.

The Little Study That Could

At first glance, Carney, Cuddy, and Yap's study appears straightforward, asking: What if a brief, deliberate change in posture could reliably stimulate a hormone profile that would in turn produce feelings of confidence and "powerful" behavior? Would those feelings and behaviors in turn result in a change of status? Though stated simply, these questions involve a complex chain of associations involving the hormones testosterone and cortisol, individual traits and behaviors, and social status, then circling back to hormones, which keeps the loop going indefinitely. "In humans and other animals," the researchers say, "testosterone levels both reflect and reinforce dispositional and situational status and dominance; internal and external cues cause testosterone to rise, increasing dominant behaviors, and these behaviors can elevate testosterone even further." Cortisol levels are a seemingly secondary part of the equation in their examples, but the authors suggest low or dropping cortisol to be another component of high status. None of this is new or controversial, basically capturing current consensus about feedback loops between hormones and dominance. Cuddy and colleagues tweak this thinking, though, by suggesting that because dominant animals use expansive body postures across species, these postures are a good cross-species cue to social status, which they also describe as power: "The proud peacock fans his tail feathers in pursuit of a mate. By galloping sideways, the cat manipulates an intruder's perception of her size. The chimpanzee, asserting his hierarchical rank, holds his breath until his chest bulges. The executive in the boardroom crests the table with his feet, fingers interlaced behind his neck, elbows pointing outward." These displays are associated with power, which the researchers define as "greater access to resources; higher levels of agency and control over a person's own body, mind, and positive feelings; enhanced cognitive function; greater willingness to engage in action; and increased risk-taking behavior." By this stage in the study's background paragraphs, dominance behaviors within any given social hierarchy have been collapsed with status. The final link in the chain of associations is hormones, which the researchers weld to some of the greatest hits in behavioral research on T, especially studies of dominance and status. Citing reviews by John Archer as well as Allan Mazur and Alan Booth, which we discussed at length in Chapter 3, the authors argue that "neuroendocrine profiles of the powerful differentiate them from the powerless," pointing to differences in both T and cortisol. This leads them to their hypothesis: since T and power rise together, it may be possible to deliberately use expansive gestures to raise T, and thereby secure an actual shift in social status.[3]

Presented as universal, the narrative from the peacock to the executive is ostensibly seamless—but let's slow down the story to look at the pieces individually. Power posing might explain why some peacocks rise to the top, but can peahens power-pose their way out of the peacock patriarchy? Did the chimpanzee and the executive get to the top by puffing out their chests or cresting the table, or is that what they do once they've acquired power? The feline example is especially strange: instead of referring to dominance behavior within a species-specific (intraspecies) hierarchy, the example is of a cross-species (interspecies) interaction. Moreover, the cat's movements seem to be less a display of dominance than of vulnerability.

Reciprocal effects between T and dominance are well documented. These complex and multidirectional effects of T are the "trouble with testosterone" that Robert Sapolsky has famously described. The long-standing assumption that animals with higher T rise to the top of dominance hierarchies gets it backward: evidence is much stronger that moving up the dominance hierarchy is what stimulates high T.[4]

In their original study, Carney and colleagues suggested something different. Instead of actually needing to upset an existing hierarchy to gain higher T, they posited, maybe you only have to adopt the gestures of the powerful. In other words, "simple behaviors, a head nod or a smile, might also cause physiological changes that activate an entire trajectory of psychological, physiological, and behavioral shifts—essentially altering the course of a person's day." Carney and colleagues were interested in exploring whether the links between power, gestures, and hormones would hold if the displays of power were posed rather than expressions of real-world power. Specifically, they wanted to find out "whether high-power poses (as opposed to low-power poses) actually produce power."[5]

Randomly assigning twenty-six women and sixteen men to either a high-power or low-power group, the researchers set out to test whether assigning subjects to a display of either dominance or submission could stimulate a corresponding neuroendocrine profile. In particular, could short power poses boost T and suppress cortisol? Moreover, they wondered whether assuming a powerful pose would make someone act more powerfully. To test this hypothesis they asked participants in each group to pose one of two ways, holding each posture for one minute, and then gave them the chance to make a bet. The choice of the gambling exercise was made based on literature suggesting certain kinds of risks are more readily accepted by people in power.

The poses were designed along two dimensions that the researchers identified as "universally linked to power: expansiveness (i.e., taking up more space or less space) and openness (i.e., keeping limbs open or closed)."[6] Aiming to capture what they thought to be common expressions of expansiveness and openness, the researchers seemed to miss the cultural specificity of their choice of a desk as a prop, which furthermore they assigned only for the high-power poses. The first high-power pose consisted of sitting tilted back in a chair with feet on the desk, hands behind head. The second was standing, with the subjects leaning on a desk with their hands. The low-power group first sat in a chair with arms tight against the body and hands folded on the lap, then stood with arms and legs tightly crossed. The researchers took saliva samples both before and after the posing to measure changes in testosterone and cortisol. To measure the effect of poses on risk tolerance, researchers then gave the participants $2, which they could either keep or wager on a double-or-nothing bet. Finally, researchers asked participants to indicate on a scale of 1 to 4 how "powerful" and "in charge" they felt.

Both the high- and low-power pose groups experienced a rise in T and a drop in cortisol, but the changes were greater in the high-power group, with similar results for men and women. High-power posers also reported greater feelings of being powerful and "in charge" and were more likely to gamble. The authors found that their study both proved that high-power poses elevate T (and decrease cortisol) and confirmed the older theory that a similar neuroendocrine change increases feelings of power and tolerance for risk. Their conclusions are dramatic:

> By simply changing physical posture, an individual prepares his or her mental and physiological systems to endure difficult and stressful situations, and perhaps to actually improve confidence and performance in situations such as interviewing for jobs, speaking in public, disagreeing with a boss, or taking potentially profitable risks. . . . Over time and in aggregate, these minimal postural changes and their outcomes potentially could improve a person's general health and well-being. This potential benefit is particularly important when considering people who are or who feel chronically powerless because of lack of resources, low hierarchical rank in an organization, or membership in a low-power social group.[7]

One might have expected this modest study to quietly fade away after publication. Instead, it was made the subject of a TED talk that launched Cuddy on a meteoric rise to fame in the public sphere that culminated in a self-help mini-industry in power posing. Designated a "highly cited paper," the original study is in the top 1 percent of psychology articles referenced, according to Web of Science.

Meanwhile, in the halls of academia, the power posing franchise was beginning to crack under the weight of withering criticism. Shortly after the article's publication, psychologist Steven Stanton argued that lumping women and men together undercut the statistical analyses: Carney and colleagues should have analyzed the men's and women's data separately, because prior data suggest not only that men and women have different T responses to dominance situations but also that they have large average differences in T levels.[8] Stanton's caution got little traction and the study authors did not reply. Meanwhile, other psychologists were trying to replicate the study's results and finding they couldn't. Using a similar experimental procedure, Eva Ranehill and colleagues at the University of Zurich studied nearly five times as many people as Carney's group but got very different results. While they were able to replicate the finding that power posing makes people feel more powerful and self-confident, they found that feeling powerful did not lead people to take greater risks, and the small effects they found on hormone levels were actually the opposite of what Carney, Cuddy, and Yap found: testosterone was higher in the low-power posing group than in the high-power posing group.[9] Their conclusion was stark: "We failed to confirm an effect of power posing on testosterone, cortisol, and financial risk taking. We did find that power posing affected self-reported feelings of power; however, this did not yield behavioral effects."

Carney and colleagues, who were invited to write a commentary in the same issue of *Psychological Science* in which Ranehill's study appeared, rejected the notion that the new study was a failed replication. Deeming it a "conceptual replication," they argued that Ranehill's study differed in key ways that might explain the discrepancy. More importantly for our purposes, Carney's group used this venue to offer up thirty-two new "allies" in the form of studies that they claimed showed the effects of power posing. Curiously, none of the studies beyond their own 2010 research and Ranehill's 2015 study include any hormone measures. Testosterone, so central to the initial hypothesis and to the narrative of the TED talk, had all but dropped out of the argument by this time. Gone was the specific hypothesis that expansive postures increase T, decrease cortisol, increase risk-taking behavior, and increase feelings of power. Instead, the studies are meant to suggest a very simple relationship: "nonverbal expansiveness" causes "embodied psychological changes." The latter are a veritable hodgepodge, ranging from increased cheating on a test and a greater number of traffic violations to improved mood and even effects on how someone estimates the weight of a box. T is gone from the mix.[10]

Now You See T, Now You Don't

Once the debate heated up between power posing's critics and defenders, T's role seemed to diminish entirely. Even Stanton's critique, which was about sex/gender differences in T, didn't pick up on all of the inconsistent data on T in the initial report. But by turning away from T, most of the narrative and virtually all of the cross-species evidence for the processes that would supposedly underlie power posing are made irrelevant.

T might have fallen by the wayside because critics focused on the more obvious problems with the statistical methods, ultimately suggesting that Carney and colleagues had effectively "cooked the books" in favor of their hypothesis. Analyzing the thirty-two studies offered up as allies, the psychologists Joe Simmons and Uri Simonsohn concluded that Carney, Cuddy, and Yap's findings were the result of "*p*-hacking." Because *p*-values are about statistical significance, low *p*-values indicate that research findings probably do not reflect chance associations. Cuddy and her colleagues had begun their analysis by first scrutinizing the data to see which analyses had *p*-values that supported their theories, and then strategically zoomed in on the variables in those analyses, while dropping other variables. Simmons and Simonsohn pointed out that *p*-hacking is part of a larger problem in science, but chided Cuddy's team for continuing to push the power posing results in the face of accumulating evidence that their results were a fluke.[11]

In 2016, in what might have been a coup de grâce against the power posing hypothesis, Dana Carney herself issued a statement disavowing it. "The evidence against the existence of power poses is undeniable," she wrote, adding that further research on the topic would be "a waste of time and resources." She confirmed that the study had been *p*-hacked. Carney's reversal laid bare how *p*-hacking unfolds in the lab when researchers are working with limited resources and are excited about their hypothesis:

using a variety of measures to capture their variables, they then selectively zoom in on certain measures only after results start to accumulate. Initially, the main hypothesis in the study was that power posing would cause someone to take greater risks in a gambling task. The idea that it would also make someone feel more powerful was tacked on later. "The self-report DV [dependent variable] was p-hacked," she confirmed, "in that many different power questions were asked and those chosen were the ones that 'worked.'"[12]

Several days later Cuddy released a vigorous rebuttal, via her publisher, advancing a revisionist version of the study's aim. In her letter, T is cut from the power posing equation once and for all: *"The key finding, the one that I would call 'the power posing effect,' is simple: adopting expansive postures causes people to feel more powerful. . . .* The other outcomes (behavior, physiology, etc.) are secondary to the key effect." The original study, as stated, was to test "whether high-power poses . . . actually produce power." Dropping T as a "key effect" dismantled the theoretical apparatus that originally justified the study, severing power posing research from the literature on animal posturing and studies of human competition. As Cuddy refigured it, power posing isn't about boosting T or even affecting behavior; it's simply about feeling powerful. From an initial hypothesis that depended on the idea of an elegant symbiotic loop among status, behavior, and physiology, Cuddy's revision pivoted to a more limited claim about behaviors and the feelings they engender.[13]

The rancor over this study transcended scientific circles, eventually featuring in a *New York Times Magazine* cover story, "When the Revolution Came for Amy Cuddy." The subheading telegraphed the gist of the article: "As a young social psychologist, [Cuddy] played by the rules and won big: an influential study, a viral TED talk, a prestigious job at Harvard. Then, suddenly, the rules changed." Cuddy, according to reporter Susan Dominus, was a "unique object" of social psychology's new appetite for critique, and the discipline's criticisms were aimed not just at Cuddy's work but at Cuddy herself: "At conferences, in classrooms and on social media, fellow academics (or commenters on their sites) have savaged not just Cuddy's work but also her career, her income, her ambition, even her intelligence, sometimes with evident malice." Dominus framed the issue as sexist, her account backed up by scores of passionate online commenters, including one who called it a "textbook case of men ganging up to bully a woman who'd achieved more success than they had." T was utterly irrelevant to the discussion.[14]

There's no question that some of the criticism of Cuddy has an ugly, sexist, and personal tone emblematic of the misogyny many women face in both social media and the sciences. But it's wrong to suggest that Cuddy's study was unassailable under some old set of rules and that she was being unfairly held to new ones after the fact. It's never been acceptable to cherry-pick variables after statistical analysis is under way. And empirical research disallows the sleight of hand by which T was first brought in and then pushed out of the story when it was convenient: anchoring the theory but jettisoning it when no empirical relationship is evident.

While Cuddy had already dropped T from the core narrative about power posing, other researchers brought T back into the limelight. The psychologists Kristopher Smith and Coren Apicella suggested that the original work was not just underpowered but flawed by its artificiality. In other words, Carney, Cuddy, and Yap's biggest mistake wasn't underpowering or p-hacking, they believed, but failing to situate the behavior they were studying within a "natural" context of competition. The backstory for power posing came from animal studies showing that powerful gestures go along with rising T and victory in real-world, often high-stakes competitions. Smith and Apicella, unlike some of the researchers who study winning, losing, and T, argued that Carney and colleagues were wrong to take competition out of the equation, insisting that the study should include a real competition, not a rigged one such as the nominal gambling exercise they chose.[15]

Smith and Apicella designed their own study involving 247 men, each of whom was paired with one other participant in a tug-of-war contest, the results of which divided the group into winners and losers. Participants in both the winning group and the losing group were then assigned to hold either a high-power or low-power pose. While winners who adopted a high-power pose did have a small rise in T relative to winners who adopted a low-power pose, T actually dropped among losers who took a high-power pose. Smith and Apicella's explanation for this result invoked evolutionary adaptations that have linked expansive postures with winning and rising T, which then encourages more winning. "To the extent that changes in testosterone modulate social behaviors adaptively, it is possible that the relative reduction in testosterone observed in losers taking high-powered poses is designed to inhibit further 'winner-like' behavior that could result in continued defeat and harm." As they figure it, power posing is no longer a way out of the lower echelons of social hierarchy. Instead, the coupling of hormones and postures has evolved to keep losers in their places.[16]

Meanwhile, Cuddy had been doing an end-run around the problem that T is an unruly hormone that doesn't behave empirically in the way that would support the claims in the 2010 study or the TED narrative. In April 2016, at the Misconceptions of the Mind Conference (MoMiCon), Cuddy unveiled a new and more explicit definition of "power." In her conference talk, "Feeling Powerless Is Not Being Powerless," Cuddy began by describing "social power, [or] power over others," before explaining that that's not what power posing is really about. Instead, she clarified, power posing addresses "personal power," which she defined as "power over our own resources like our knowledge, our core values . . . the list goes on and on." She never articulates that this sort of power is different from the type of power she was describing in the TED talk or the study. Instead, she speaks of being affected by the many people who contacted her after the TED talk to describe their own experiences with power posing. What mattered most to them, they said, was not any particular outcome, like getting a job or acing a test, but how they felt after tackling a challenge. The people who felt good were the ones who had presented their "authentic selves," Cuddy explained.[17]

Moving briskly from her explanation of personal power, she offered a minor remix of the images and talking points from her TED talk: "Power causes us to expand in many different ways, and this is related to the approach system. We open up. It does the same thing with other animals as well. So when individuals have power, they expand, they make themselves bigger, they take up space." Behind her, we see the by now familiar images of a chimpanzee, a gorilla, wild canines, swans, and a peacock from her TED talk. Most viewers watching both talks won't immediately spot a problem, but it's there: while Cuddy might be talking about "personal power," research connecting expansive posturing across species is about social power, or rank in hierarchies. T, of course, is an integral part of that research, which means that T hasn't really exited Cuddy's frame; it's just been made into backstory.

Whether or not Smith and Apicella are correct to suggest that winning and losing must be earned (or "naturalistic," in their language) to evoke the relationships among competition, postures, and hormones, they do hold themselves accountable to the larger literature on expansive postures and power by retaining T as a key variable. In Cuddy's version, T doesn't need to be measured, the result being that it can't disrupt the story if and when it doesn't act according to expectations. Personal power may indeed be interesting and important, but you can't have it both ways. You can't use that concept of power to situate power posing in the larger literature on how animals with "power" take up space.

There is a delicious irony to the idea that T—of all things—might be capable of propelling women, or anyone else in a "low-power social group," out of a subordinate position. But T turns out to be an unreliable ally in this fight. Data from multiple fields strongly suggest that while T is responsive to social situations and our physical exertions, it's not a simple switch that we can turn on and off. None of the previous critiques offered a close examination of how T was supposed to work in this process. Maybe it's because the T part seemed plausible. Maybe the critics didn't understand the literature on T well enough to go after that part of the story. Maybe it's because there was so much low-hanging fruit in the study (such as the *p*-hacking) that nobody got around to looking closely at how T fit into the model. We can't say why it happened, but for all the voluminous critique, one of the core T narratives—"power is about T"—emerged unscathed from the ashes of the debate.

There is another lesson here beyond T and power, though. It is concerned with how theories can be recuperated by dropping data, and data recuperated by shifting theories. What do you see if you keep your eye on T? For one thing, the initial Carney, Cuddy, and Yap study would have looked implausible sooner rather than later. You don't have to do a fancy *p*-curve analysis to see that the T data just don't add up. As Ranehill and colleagues point out in a letter posted on the blog Data Colada, Carney and colleagues' descriptions of T in the text of their original study indicate an average *decrease* in T, whereas in a key figure, the numbers suggest that T *increased* in the sample. Moreover, none of the studies that Carney and colleagues later marshaled to support the effect of power posing even report T values, minimizing their impact on the claim. Still, in the final point of that review, Carney, Cuddy, and Yap returned to T to anchor power posing to a larger literature. They treat T as a malleable explanatory resource to be drawn on or omitted, depending on whether it supports their narrative.[18]

Keeping your eye on the T also brings gender to the foreground. Early on, Steven Stanton wrote the only published critique that focused on how T as a variable was modeled in Carney and colleagues' 2010 paper. Both of Stanton's chief complaints had to do with gender. First, reasoning that the link between powerful poses and T is about dominance, Stanton says that evidence on T and dominance suggests T rises when men win a contest, but there's no similarly observable T effect in women. Thus, Stanton argues that Carney, Cuddy, and Yap should have done separate analyses by gender. His second critique is more basic in nature, and thereby less dependent on interpretations of the broader literature on T and dominance. Homing in on the apples-and-oranges quality of the dataset, he points out that T is not a normally distributed variable in a combined sample of men and women. Because men have, on average, much higher T levels than women, a graph of T scores has two distinct peaks instead of a single bell curve. Because the researchers analyzed their data as if T were normally distributed, their results might be an artifact of incorrectly pooling data from women and men.[19]

Paying close attention to T also returns us to the assumptions about power relations that underwrite the various studies. Carney, Cuddy, and Yap's model is built on the familiar volumetric story about T: more T is linked to more power. Once you accept this premise, everyone wants in; moreover, everyone sees themselves as deserving to be in. But resources remain limited. Within these parameters, you can grab more power, women can move up the ranks, but the natural order of things is always going to be a hierarchy where men outrank women, based on their naturally higher average T levels. This isn't so different, in the end, from Smith and Apicella's model. There, power is an ontological property of "winners" and "losers," and the bedrock of power is physical competition. Their focus on physical competition makes a lot of sense if the goal is to tie the human studies to the animal research. But they also link to Mazur and Booth's biosocial model of status, which holds that dominance in humans usually is not physical, but is about human status hierarchies and resource accumulation. Either way, if you follow the implicit scientific theory of power out far enough, the connecting threads start to fray.

T for Ladies

Carney and colleagues crafted their hypothesis by combining two different lines of evidence: first, that shifts in social status stimulate shifts in endocrine profile (T rises, cortisol drops); second, that higher-status animals use expansive gestures. Whether faking a status change by adopting the gestures of the powerful could also stimulate a shift in endocrine profile would have been a plausible question. But they made two general kinds of errors, and critics have only focused on the first kind, which is in some ways more trivial. Those errors all fall under the heading of questionable scientific practices that made it look like their hypothesis was supported when it wasn't—*p*-hacking, errors in treating T as if the different distributions in women and men didn't matter, and so on. These errors can be fixed with better statistical practices.

The second type of error they made is even more fundamental, because it concerns their broad conceptual model. Even if they had legitimately found that expansive gestures boosted participants' T levels and lowered their cortisol, and even if the power poses had caused participants to take more risks in the gambling tasks, it still would have been a profound error to suggest that those effects would smoothly translate to shifts in real-world social power.

When the power posing controversy is presented as mainly a case of bad scientific practice or bullying a woman scientist, it obscures the underlying reasons behind power posing's broad appeal, why it rose to prominence so quickly, and why it still has legs in spite of the scientific critiques. The power posing phenomenon emerged amid a decades-long national conversation about women's continuing secondary status in spite of concerted policy efforts to close these gaps. Power posing has a progressive flavor: women should be able to exercise power and feel powerful, even the kind of power that's underwritten by a burst of testosterone. It is the biopsychosocial version of Sheryl Sandberg's bestselling book *Lean In: Women, Work, and the Will to Lead*. Sandberg and her book's associated movement want women to change their behavior; Cuddy says power posing will help them to make those changes. Sandberg wants women to lean in; Cuddy wants them to manspread. Sandberg even makes a direct appeal to power posing, writing, "Research backs up this 'fake it till you feel it' strategy. One study found that . . . [a] simple change in posture led to a significant change in attitude." But as Michelle Obama recently noted when on tour for her memoir, *Becoming,* "It's not always enough to lean in because that shit doesn't work."[20]

As with the studies of aggression and T that suggest patterns of violence and criminality can be explained by too much T, the power posing narrative uses T talk to shrink a huge social problem to the microenvironment of an individual body. If power posing can, as Cuddy says, "significantly change the way your life unfolds," why bother directly confronting structural inequality? Standing with your hands on your hips provides a fast, individualized solution that is much more manageable than challenging entrenched systems. It's the kind of argument that appeals to a liberal feminist sensibility, even as it obscures the radically different relationships to power among the homogenized category of "women." In spite of the feminist sheen, the power posing approach actually counters a mainstay of feminist analysis and activism: gender inequality stems from social formations, not from biology. For those researching power posing, power is internal: postures shift state of mind and hormones, which in turn shift behaviors, creating power itself.

Power posing is simultaneously a call to female empowerment and a reflection of the long-standing quest for self-improvement that has had particular potency in the US context. Citing fellow historian David Serlin, Evelynn Hammonds and Rebecca Herzig document that hormonal self-improvement is one of the early ways in which "medical reinvention came to be seen as a purchasable amenity. Like a new frost-free refrigerator, hormonal intervention might become a commodity like any other, a tool for remaking the protean American self."[21]

It's not just that hormones can change, but that women should change them. Plasticity plays a key role in contemporary imperatives of biological citizenship, and taking charge of one's hormones is merely one example of how a person might transform herself and thus her life. Power posing is what a responsible woman who wants to be effective in the workplace or the world more generally should do. This echoes a process that has been at work in the popularization of other scientific fields, especially neuroscience with its double-edged concept of brain plasticity. As sociologist Victoria Pitts-Taylor has remarked, the discourse of plasticity "opens up the brain to personal techniques of enhancement and risk avoidance . . . where the engineering and modification of biological life is positioned as essential to selfhood and citizenship." Following an era of great expansion in state regulations meant to safeguard citizens' health and well-being—as in regulation of environmental toxins, workplace regulations, and structures of oversight for pharmaceutical development—more responsibility has devolved to individuals. Likewise, the post–World War II decades in which an official ethos of corporate responsibility reigned have given way to an era of the lean, dynamic corporation whose responsibilities are not to workers or communities but to shareholders above all. Flexibility means workers have to be more available, work harder and smarter, and still understand that their employment is "at will" rather than guaranteed, regardless of their performance. This lack of security is also framed as opportunity: workers have to be prepared to reimagine themselves, and the very best ones will simply start their own games. Would that power posing could help with that, too.[22]

In arguments about human differences, Hammonds and Herzig show that hormones have long been "used in competing ways: alternately to affirm absolute typological distinctions between bodies or to demonstrate the plasticity and continuity of an ambiguous spectrum." If T leads to risk-taking and risk-taking is how one gets to the top, then men (with their higher T levels) are the ones destined to be on top, while women's natural place is lower down the ladder. But since T is malleable, then the gender hierarchy might be mutable, too. Echoing what the anthropologist Emilia Sanabria has documented in the realm of pharmaceutical use of hormones, the solidly masculine identity of T might be primed for a correction. Sanabria shows that in some circumstances, for some ends, women are eager to take T because they (and their doctors) are convinced that it will extend some of the pleasures and benefits of masculinity to them without undermining their femininity: "Androgens (such as testosterone) may now be summoned with no apparent contradiction for patients or doctors in the making of new forms of femininity. With testosterone, women can be like men and yet remain women, we are told. They can become super-women." Sanabria has described the prescription of so-called sex hormones, including testosterone for cisgender women and estrogens for trans women, as "exogenous sex."[23]

Whether the focus is on endogenous or exogenous T, it seems like the long-standing conflation of T with masculinity is under pressure, which might introduce new elements into T's multiplicities and generate new questions. Is "T for ladies" about women leveraging a general resource—which thereby positions T as universal and disrupts its identity as a male sex hormone? Or when women boost their T, are they instead appropriating a little bit of maleness, which recapitulates the sex hormone identity? While boosting T pharmacologically could be viewed either way, the idea that women's bodies are capable of responding to particular interventions by producing more T challenges the sex hormone concept, converging with critiques advanced by feminist and queer scholars: calling T "the male hormone" is more about power hierarchies and gender ideology than it is about physiology or biochemistry.[24]

Breaking masculinity's iron grip on T may seem like it could denaturalize male power, too. While Sanabria shows that there's room for T within a form of feminine gender, Paul Preciado, a theorist of gender and sexuality whose book *Testo Junkie* documented a year of auto-experimentation with T, argues that T can be (and is) used to "hack" gender. When T is deliberately brought high in people who should supposedly have low T, such as people designated female at birth who want some of the masculinizing effects of T but don't want to transition or go through a medically regulated protocol, gender is destabilized. Preciado suggests that appropriating pharmaceutical T outside of official medical practices "shift[s] the code [of gender] to open the political practice to multiple possibilities."[25]

The idea behind power posing has a potentially even more revolutionary message: T is *already* as much feminine as masculine, a resource that anyone can call on from within their own bodies. The fantasy version of power posing thus queers T and makes endogenous T as well as pharmaceutical T into a technology with the potential to elicit and legitimize new arrangements and relationships between the categories of "masculinity" and "femininity," between social status and power, and between bodies themselves. But this fantasy requires a predictable chain of reactions—from a powerful pose to a rise in T to a range of effects that this rise would supposedly have—and that's not the way T works. T is responsive, but it's not a simple switch. Moreover, the last step in the chain, where elevated T supposedly cements one's powerful position, is broadly inconsistent with human evidence.

Does this thought experiment offer any insights for the way that exogenous T is used to hack gender? Perhaps. It's relatively straightforward to use T to stimulate some changes on the body's surface that will be read as masculinizing: increases in facial and body hair, change in skin texture, a receding hairline, a roughening or deepening of the voice. In this way, exogenous T is a technology that can be reliably used to hack gender. But does it hack power? Insofar as people can gain power by being read as more masculine, this shifts who has power, but it doesn't change the power system. The only reliable path for gaining power from T is to be read as more masculine: most of the knowledge about T's material effects suggests that it is unlikely that T has meaningful direct effects on people's behaviors and psychological traits.

• • •

WHILE POWER POSING ADVOCATES don't suggest that people literally "do the Wonder Woman" in front of the boss while negotiating a salary, adopting the pose in the bathroom is supposed to make you feel and act more powerful out in the world. Carney, Cuddy, and Yap suggest that power posing may be especially useful for those with the fewest resources, presupposing that power posing works similarly for everyone. The theory of power posing presumes generic individuals with no particular racial identity, sexuality, religion, gender presentation, class, education, ability, or body type. The power each individual wields is considered transferable, fungible, not connected in any meaningful way to history, broader social arrangements, or the specific contexts of "powerful" gestures or behaviors. But *which* woman in *which* setting alters meaning: a community meeting, a hallway at an investment firm, and a queer service organization all have different codes for what power looks like, who is supposed to wield it, and what constitutes transgression. There is no simple formula that maps postures and power onto bodies. Moreover, oppressions aren't additive, they are qualitatively different: specific social rules and the consequences for breaking them are dynamically produced by interactions among multiple axes of power.[26]

Joan Williams, a professor at the University of California Hastings College of the Law, studies biases against women of color in the traditionally male-dominated STEM fields, who must "walk a tightrope" between appearing too feminine to be competent and too masculine to be likable. Williams's data demonstrates how gender stereotypes are racialized. Among Asian women she interviewed, nearly half reported feeling pressure to behave in "feminine ways," while just 8 percent of black women said they did. When women's behaviors challenge expectations of femininity, they trigger racial stereotypes, such as the trope of the "fiery Latina" as "hot-blooded," "irrational," "crazy," or "too emotional." Sixty percent of Latinas she interviewed said they experienced backlash when they expressed anger or weren't deferential. Power posing, not to mention the resulting "bold" behaviors Cuddy cites, could hinder more than help their position.[27]

When people in lower-status positions act powerful, they also confront the social rules of hierarchy, and risk being labeled as difficult, confrontational, negative, aggressive, or worse. It's not just that power posing can't fix the systemic roots of inequality. It's that power posing can be counterproductive or even dangerous for people who are viewed as a threat to the systems they are trying to enter or change. Take, for example, US football quarterback Colin Kaepernick, who sued the National Football League for colluding to keep him off the field because of his peaceful protests of police violence against people of color. His protests, which involved silently taking a knee while the national anthem was played before games, caused an international stir and backlash that played out on television, in boardrooms, in courts, and on the president's Twitter feed. There are two takeaways from this example. First, power doesn't inhere in specific postures: Kaepernick's kneeling pose would in the abstract be identified as non-threatening and even submissive, but in the context of the political commitments that motivated the pose, it was viewed as challenging in the extreme. Second, that challenge didn't propel Kaepernick to a more powerful position, but on the contrary left him locked out of professional football, targeted by ridicule, protests, and presidential threats.

Social hierarchies aren't passive phenomena: they are constructed and enforced. Responding online to a public radio show about power posing, Beverly Smith, a black feminist health advocate and writer who is also disabled, asked "How does this work for Black people? What about Black women like Sandra Bland and me?"[28] Smith was referring to the 2015 death of Sandra Bland, a twenty-eight-year-old woman who died in police custody, allegedly by hanging herself with a garbage bin liner in a Texas jail cell. Three days earlier, Bland had been pulled over in her car by state trooper Brian Encinia, who said she had changed lanes without signaling. During their exchange, some but not all of which was captured by Encinia's dashboard video camera, Bland repeatedly asserted her rights by questioning the basis for the officer's demands. As a writer for *Huffington Post* observed, "A close look at the police car dashcam video that recorded the exchange shows her questions had merit: Encinia at every occasion escalates the tension. He tells Bland, a Black Lives Matter activist, she's under arrest before she has even left her car, shouts at her for moving after ordering her to move, refuses to answer questions about why she's being arrested and, out of the camera's view, apparently slams her to the ground. He gets testy with her—'Are you done?'—when she explains after he points out she seems irritated. And, contrary to a recent Supreme Court decision, he unconstitutionally extends the traffic stop, it appears, out of spite."[29]

A pivotal moment in the power dynamics between the two is discernible: when Bland is still seated in her car, Encinia asks her to extinguish her cigarette, and she (correctly) states that she is not obligated to do so because she is in her own vehicle. Encinia's demeanor shifts, and from this point he is visibly angry, ordering Bland to get out of the car. When she continues to sit in the car, Encinia reaches in and slaps her, pulls out a Taser, and tells her, "I will light you up." Bland then gets out of the car on her own, but Encinia directs her outside the dashcam's area of focus. Twelve minutes later we can hear Bland saying repeatedly, "You're about to break my wrist," and then screaming, "Stop!" Encinia later charged Bland with assaulting a public servant and took her to jail, where she was put in solitary confinement. Seventy-two hours later she was dead.

Did Sandra Bland hang herself or was she killed? The police claimed that she died by her own hand, but autopsy results were inconclusive. Beginning with friends and relatives back in Chicago, who were shocked by the rapid sequence of events and didn't believe she would have killed herself, protests were organized and people around the country questioned the circumstances of her death. "'A minor traffic infraction should not turn into death,' said an organizer at a protest at the [Waller County] Courthouse." Echoing Beverly Smith's concern about the risks for black women of being assertive (in Bland's case, merely vocally so), one protestor held a sign that read: "I am an outspoken Black female who knows my rights!! It could happen to me. #IamSandraBland."[30]

Power posing reinforces widespread simplifications made about human power relations. By saying this, we are not implying that those who support power posing would intentionally advocate something harmful. But by failing to acknowledge that individual people sometimes have little room to maneuver and may face dire consequences when they try, this radically individualized model of power reproduces hierarchies by denying their existence.

As far as the science goes, the theory of posing your way to power is dead. But in the broader culture, power posing is still among the undead: Cuddy's TED talk continued to amass millions more views even after it was discredited, and it's still going strong. Cuddy's power posing research has all the ingredients for making a zombie fact: strong resonance between a scientific finding and familiar cultural stories, timed precisely at a moment when this new finding seems to answer an important social question (in this case, what to do about power imbalances throughout the culture). T comes and goes, but that doesn't seem to matter, perhaps because this phenomenon was never about the molecule T in the first place. If T is instead read as a metonym for power itself, the power posing narrative makes more sense. This is a great example of the way triangulating with T works: if T is male and power is male, then T links up easily with power, whether or not there exists a direct relationship between the two.

Power posing research began as a node in a scientific network about social hierarchies and the embodied effects of change within those hierarchies. But there's a problem with the theoretical links in that network: once the researchers zoomed in on the power of individuals, the hierarchy itself faded away. In the revised framework, power moves resemble one player moving pieces on an empty chessboard with no opponent. There's no opportunity for relational effects, certainly no possibility of backlash. Social hierarchies, though, are dynamic systems: when one person moves up, there's a reaction from the people who used to be at the top. There's also a data problem with the revised model: the power posing studies about the effects of expansiveness show no consistent effects on T. The model was supposed to be about "embodiment" processes, meaning the concrete mechanisms by which power and bodies are connected, but in the end it has no internal bodily components; it's a simple assertion about the connection between surface gestures and affect.

The story of power posing can tell us a lot about the curious career of popular science and the work that T does, nimbly shifting from main character to part of the unnamed chorus, waiting offstage until the right moment to make another supportive cameo. Power posing is still chugging along, still on the TED website, with no disclaimer. In this social realm, where T's many guises include masculinity, power, biology, and science, T's initial presence in the power posing story gives "truthiness" to the cultural narrative that power flows from T and that social hierarchies are natural. In the scientific realm, where T is meant literally as a hormone in a biological process that involves feedback loops of neural signaling, postures, social feedback, and so on, the revamped power posing hypothesis (without T) is no longer connected to the theoretical literature in which T is central. But once T is part of the story of power, it can't simply be cut out, because the scientific and social realms act simultaneously. Elements can drop in and out of the theories, definitions can switch, evidence can be lacking, but it's very hard to follow all the threads, and in the end it doesn't seem to matter much that the science of T no longer supports the power posing narrative.

The fantasy version of power posing, where we can call on flexible biologies to reshape power structures, seems very different from old-school biologization, where stable biologies ensure that social structures remain locked into place. Traditional T was masculine, driving men in the fast lane down a one-way highway of power. The power posing story capitalizes on a newer understanding of T as socially responsive. If this flexible

T can be harnessed, it seems to democratize power: it's a free resource that even the most powerless people can call on to change their social positions. Maybe it's even queer, eroding boundaries between masculinity and femininity and shifting relations between bodies, hormones, and social status. In contemporary scientific and social narratives, the body is no longer a strictly bounded individual but is porous. Social context enters and transforms bodies, as when social status and hormones shift in tandem. But social formations are not just scaled-up versions of the interactions between an individual body and the environment. From the glossiest pop version of T science that's found in the TED talk on power posing to the more sober and scientifically careful versions that actually track T in bodies, the nature of power is obfuscated by ignoring how power is reproduced, how it is connected to the material world outside bodies, and more. The concern for persistent inequalities that seems to have set the stage for power posing's rise can't be addressed by an intervention or a science that rests on a version of power that's isolated from large-scale social processes and material underpinnings. If there's one lesson to take away from the strange and strangely undead idea that T will help you pose your way to power, it is that "flexible biologies" are no better for understanding and (re)shaping the social world than essentialist biologies were.

5

RISK-TAKING

O N HER BIRTHDAY in the fall of 1901, Annie Edson Taylor rowed out to the middle of the Niagara River and contemplated the spectacularly raging Horseshoe Falls, with its infamous vertical drop of 167 feet. With the help of two men, she strapped herself into a wooden pickle barrel weighted with a 200-pound anvil and lined with a few cushions. She placed a plastic air tube between her lips while the barrel was pressurized with a bicycle pump and then sealed. At sixty-three, in a black silk dress and a jeweled choker circling her throat, hair upswept and topped with a single ostrich feather, Taylor looked decidedly more elegant than intrepid.[1]

At 4:05 P.M., Taylor was released into the Niagara River and immediately swept up in what the *New York Times* called a "frightfully swift" current: 5 billion gallons of water per hour flowed over these falls to Grass Island below. During her mile-long trip down the river to the lip of the falls Taylor submerged out of sight several times but always reemerged. By 4:23 P.M. she tumbled over the edge of the river and into the cascade. She hit bottom in less than a minute, the first person to make it over Niagara Falls in a barrel and survive. She bobbed in the eddies for another fifteen minutes before being rescued, but eventually peeked out over the lip of the barrel, its straps likely having saved her neck from breaking from the impact of the fall. The next day's *New York Times* trumpeted: "Woman Goes over Niagara in a Barrel. She Is Alive, but Suffering Greatly from Shock."

When asked for a public comment about the experience, she addressed her inevitable would-be emulators: "Don't try it."[2]

• • •

TAYLOR WAS BORN IN Auburn, New York, into a well-to-do family. It was during a four-year training course in Charlottesville to become a teacher that, aged eighteen, she met and soon married David Taylor. After almost twenty years of marriage, David was killed in the Civil War, launching Taylor into a peripatetic existence as a widow plagued by financial troubles. Accustomed to a certain lifestyle, and not making much money herself, she relied for a time on an inheritance from her parents. By 1898 she had crossed the continent eight times looking for employment as a dance instructor, eventually settling in the small town of Bay City, Michigan, to found her own school. It didn't provide her the money she needed, so she went to Texas and Mexico City to find work, but there was never enough.

By 1901 Annie Taylor was back in Bay City, impoverished and living in a boardinghouse, when an article about the Pan-American Exposition in Buffalo describing two huge waterfalls in upstate New York caught her eye. "I laid the paper down," she said, and "sat thinking, when the thought came to me like a flash of light—'Go over Niagara Falls in a barrel. No one had ever accomplished this feat.'" Though the accomplishment brought Taylor notoriety, her Niagara plunge was not lucrative: she worked as a Niagara street vendor for twenty years and died penniless in 1921.[3]

• • •

HIGHLY RISKY BEHAVIOR LIKE getting into a barrel and hurling oneself over one of the world's largest waterfalls is not easy to study. But there is a plethora of research that does look at risk-taking, and much of it seeks to link this behavior to T. By this point in the book, the core hypothesis of these studies will be painfully familiar: men take more risks because they have higher T levels on average than women, and higher T primes them for risk-taking behavior. So it has been since time immemorial, goes the presumption, because T, that manly hormone, was paired in early human evolution with traits and behaviors that gave high-T men a survival and reproductive advantage.

You might think we began with the story of Annie Taylor to argue that the great women daredevils have been overlooked by history, or more broadly that women take risks, too. But that's not our point. We are interested in Taylor's story because it points away from T, instead highlighting how gender, poverty, and other forms of social status interact to structure someone's life chances and choices, and the risks and benefits to that person of different actions. Annie Taylor may have been singularly daring and creative in dreaming that her way out of poverty might be to cram herself into a barrel and plunge over one of the world's largest waterfalls, but desperation can make people take dramatic risks.

Risky Business

Risk-taking is part of a suite of behaviors that are linked to T through popular versions of the scientific theory of sexual selection. T is the proximate mechanism that supposedly ensures that males compete, take risks, want lots of sex, and generally fulfill their evolutionary destiny. In her recent book *Testosterone Rex,* the psychologist Cordelia Fine skewers the idea that modern-day sex differences in behavior and circumstance mostly reflect a deep evolutionary past that favored bold, competitive behavior in men and cautious, nurturing behavior in women. The long-standing belief is that reproductive success, commonly measured by how many offspring live long enough to reproduce, is not an even game for females and males. Women, who bear a greater physical burden in reproduction, can produce only a fraction of the offspring that are possible for men. This limited number of potential offspring requires women to be cautious and choosy about mates, prioritizing quality over quantity. For men, it's a straightforward numbers game: because the cost of reproduction is low, the more partners and the more encounters the better. But for some men to reach the glorious achievement of hundreds of offspring, they have to monopolize the women available to them, which means some men will not reproduce at all. Men have to compete among themselves to be lucky enough to reproduce, while women don't have this pressure. According to this storyline, men evolved to be risk-taking, promiscuous, and competitive, while women developed an inclination "to play a safer game, more focused on tending to their precious offspring than diverting their energy toward chasing multiple lovers, riches, and glory." Fine reveals fatal flaws in the human and nonhuman evidence for this theory, leaving it full of holes. But her analysis also suggests just how many hurdles you have to jump over to pry apart risk-taking from masculinity.[4]

Once again, T plays a star role in the grand story that links evolution to modern-day gender differences. T, in this schema, is responsible for the whole package of physiological, psychological, and behavioral masculinity, so risk-taking must be part of T's portfolio. But "must be" and "is" are two different things.

•   •   •

A GLANCE AT THE LITERATURE shows how the notion of risk covers phenomena both weirdly narrow and wildly divergent, ranging from riding a motorcycle without a helmet to risking a week's earnings on a poker game. It's hard to miss, however, that many of the studies, especially the most frequently cited ones, involve money. Researchers study hormones and behavior among entrepreneurs and traders, and they use a wide array of gambling games to examine financial risk-taking, mostly among students.[5] Risk-taking and business show up hand in hand as cultural touchstones in web searches that yield results like "5 Things the Smartest Leaders Know about Risk-taking," and in online definitions from BusinessDictionary.com and Forbes.com. Even the Merriam-Webster dictionary provides this sole example of the term "risk-taking": "Starting a business always involves some risk-taking." In keeping with this cultural habit of imagining risk as risking money, most researchers studying risk-taking treat financial behavior as the epitome of risk-taking in humans. Of course, this skips over obvious problems, such as that you have to have money to risk money, and that who has money is determined by all kinds of external factors, like gender systems, social class, and global economic relations, as well as idiosyncratic life events. In this chapter, we home in on studies that link T with financial risk to highlight how they are propelled by assumptions about class as well as gender, and to show how the studies wobble when these assumptions are brought into the open.[6]

Entrepreneurs

One way that financial risk-taking gets connected with T is to explicitly extend the familiar evolutionary tale about men and risk-taking to include "business behavior" as one of the activities that confers reproductive ad-

vantage. The psychologist Coren Apicella and her colleagues seem to position money and finance—indeed, all resource acquisition—as the
rightful evolutionary heritage of men: "Monetary transactions are a recent phenomenon in human history, but the acquisition and accumulation of resources by men is not. Money is, in this sense, a proximal currency used to maximize returns in some other currency, such as utility or
fitness. Men may have evolved to engage in riskier behaviors compared
to women because the potential returns in terms of fitness payoffs can be
higher."[7]

Similarly, in a widely referenced 2006 article, Roderick White, a management professor, teamed up with two evolutionary psychologists,
Stewart Thornhill and Elizabeth Hampson, to use "evolutionary psychology as the theoretical perspective for exploring the relationship between a heritable biological characteristic (testosterone level) and an
important business behavior (new venture creation)." In a study of MBA
students, they looked for correlations among a measure of risk propensity, a single measure of salivary T, and responses to a question about experience as an "entrepreneur." They found positive relationships between
T and both risk propensity and being an entrepreneur, concluding that
"new venture creation is more likely among those individuals having a
higher testosterone level in combination with a family business background." White and colleagues' debt to evolutionary psychology is
opaque in their hypotheses but shines through in their literature review,
which reads like the greatest hits of studies on sexual selection, testosterone, and contemporary sex differences in behavior. They had to do a
little work to fit entrepreneurs into the classic evolutionary tale of how
males had to risk their lives to reach the top of the social ladder, hurting
or even killing each other over access to women. But they find the thread.
"In our ancestral environment," they write, specifically pointing to the
Pliocene epoch, aspiring alpha males needed "initiative, persistence, and
assertiveness" to challenge "the incumbent." "By definition," they say, "entrepreneurs engage in a risky behavior (new venture creation) and are generally believed to have a higher psychological propensity towards risk
than non-entrepreneurs." To expand the landscape of empirical support
for this idea, they pair behavioral research on T with entrepreneurship
studies, noting that "risk is a central concept in both literatures."[8]

But what, exactly, is the concept of risk that runs through these literatures? Leaning hard on previous studies by James Dabbs, especially the
study of army veterans that Dabbs used to link higher T to lower social

class, White and colleagues assert that "the preponderance of evidence supports a relationship between T and occupation." Their breezy tour through Dabbs's findings reasserts the idea that higher T leads to "blue collar" occupations, while lower T leads to "white collar professions." White and colleagues also highlight some interesting occupational contrasts from that study, such as the "male actors and professional athletes" who have higher T levels compared to "ministers and farmers." We were intrigued by these groupings, and decided to go back to the study they came from. It turns out Dabbs and colleagues were also initially unsure what to make of them, writing, "We do not know how actors and football players are similar to one another and different from ministers. Nor do we know why the assertiveness of salesmen, the high status of physicians, and the sensation-seeking qualities of firemen were not reflected in higher testosterone." As usual, though, these puzzling results were no match for Dabbs and his team, who quickly provided several possible explanations for their "negative findings." Whenever men in characteristically aggressive, high-status, and sensation-seeking jobs had relatively low T, the researchers rationalized the illogical results. The salesmen in their sample, for example, "were not involved in aggressive 'cold call' selling, the physicians still had resident status, and the 'sensation-seeking' firemen spent more time waiting than fighting fires." Why bother digging up lapses in logic from a study that's three decades old? Because it is providing ballast for a much newer and frequently cited study.[9]

There's no clear concept of risk running through Dabbs's studies on occupation and T, but what about the concept of risk in White's study of entrepreneurs? White and colleagues defined entrepreneurial behavior as "full-time involvement in creating a new venture," a measure they say "is frequently used as a simple, functional, operational measure." They studied first-year MBA students, classifying them as "entrepreneurs" or "non-entrepreneurs," according to the students' self-identification as having been "involved in a new venture start-up prior to their MBA studies." The researchers then scrutinized the students' descriptions of their particular involvement in the start-up to further distinguish real entrepreneurs from what they called "possible" and "non-" entrepreneurs. At first glance, the procedure looks entirely reasonable: they eliminated those who "did not have significant full-time involvement in the new venture; often they were part-time employees, passive investors, or board members." But on closer inspection, the sorting method begs some of the most central questions about the degree of risk involved for the various participants. Who risks more, a full-time employee who sinks $10,000 into a project but has a nest egg of $50,000, or a part-time employee who sinks his entire life savings of $10,000? How "new" does the new venture need to be? If your family is in real estate and you draw on their contacts to create a venture in a neighborhood they haven't yet invested in, is that a start-up? Is that entrepreneur as risk-tolerant as the passive investor who decides to mortgage her own house to back a brilliant new tech idea? Do these differences matter?

Let's step back to view the bigger picture. Risk tolerance and risk aversion are two characteristics that supposedly connect T level to occupation, but another one is status. Remember the evolutionary story that White offered: Pliocene-epoch human ancestors, at least the males who merited theorizing about, achieved status by being risk-takers. Today, in the modern era, T is the mechanism that keeps men taking risks because, by White's determination, "evolution rarely discards anything."[10]

Dabbs's research is again White's go-to for evidence on status and T, and in the largest study available, with over 4,000 men, Dabbs found an inverse relation: higher T correlates with lower status. Not surprisingly, there's a problem with the causal chain here. In Dabbs's study, unemployed men unequivocally have the highest T, followed by lower-status blue-collar workers; white-collar workers measure third-highest, and farmers place in a pitiful fourth position. White's theory, in contrast, is that higher T is the result of ancestral male behavior: men taking great risks in order to earn high status, amassing enough resources to reproduce more successfully than lower-T, less daring males. The studies claim a shared theoretical basis, but this particular theory cannot do double duty and support them both simultaneously.

Nonetheless, by focusing on risk-taking instead of status, White recruits Dabbs's work for a narrative that fundamentally suggests titans of industry are where they are in part because of their "risk-taking propensity," which in turn flows from the heritable trait of innately higher T. In fact, the largest study of financial risk-taking that we have seen shows "no evidence for any biological effects." In a study aimed at disentangling biological and environmental contributions to risk-taking in investment behaviors, where risk is measured by stock market participation, the investigators used data from 2 million Swedish men and women who were raised by their biological parents and 3,275 adoptees for whom demographic and financial data were available for both biological and adoptive parents. When looking at financial behavior in the full sample, they found that about two-thirds of the variation in investment behavior between the adoptees versus the non-adoptees was attributable to characteristics of the adoptive family, which left a sizable contribution from genetics. But they saw a problem: the complete dataset did not screen out people without any wealth to invest. Obviously, for those individuals, failing to invest in stocks isn't an indicator of their "true risk preferences." By restricting their analysis to people who had assets to invest, they concluded that "intergenerational correlations in the stock shares are only evident for adoptive parents [and] suggests that risk attitudes may be environmentally rather than genetically determined." More bluntly, "once one controls for parents and children having positive financial wealth, there is no evidence for any biological effects."[11]

Iowa Gambling Task

One way to study risk-taking is to select real-world behaviors and compare people who seem to make different decisions about risks. That's what the studies of entrepreneurship are trying to do. But the major shortcoming of those studies is that subjects' social and economic circumstances, including both structures and random opportunities, can never be fully known or measured. Investigators might overlook some important differences between those who do and don't take certain kinds of risks, as the study of Swedes who do and don't have wealth to invest demonstrated.

Laboratory studies of risk get around this problem by presenting every study subject with an identical set of resources to put at risk on an identical set of actions. The Iowa Gambling Task (IGT), used in several influential studies of risk-taking and T, is a great example of this.[12] Each participant is given four decks of cards and an equal amount of cash, and is told to play to win the most money. They turn cards over one at a time, and each card carries either a reward or a penalty. The decks are loaded so that two of the decks give higher immediate rewards but also have more penalty cards. The winning strategy over the long run is to choose the decks with more modest rewards but also fewer penalties. To date, six studies of T and risk-taking use the IGT.[13]

The way in which the IGT links risk with T is surprising in that it suggests that higher T is connected with irrational and even pathological decision-making patterns. The IGT was originally developed to assess decision-making and emotional processing in patients with brain damage and has also been applied to study people with psychiatric diagnoses, such as schizophrenia and drug addiction. The test is seen as useful because it differentiates between "healthy" and "pathological" decision-making processes: choosing the longer-term strategy to make money is considered healthy, whereas choosing greater short-term rewards, with their associated larger penalties, is considered unhealthy. Here, some of the usual elements of the notion of risk such as novelty and sensation-seeking fall away; "risk" is conceived as a faulty risk-benefit calculation, an irrational attachment to the thrill of a big win, and insensitivity to the punishments that go along with those wins. Viewed through the IGT, risk-taking isn't about boldness that can increase your status and ultimately increase the resources you can devote to your reproductive success. Instead, it's a decision-making deficit that will cause you to end up at the bottom of the heap. If higher levels of T lead to "riskier" strategies on the IGT, then a problem emerges within the bigger-picture evolutionary theory that is supposed to tie it all together.

The researchers don't seem to notice this problem. Jack van Honk and colleagues at Utrecht University in the Netherlands were the first to use the IGT to explore the relationship between T and sensitivity to punishment and reward. Situating their study in the literature on T and dominance-related behavior, van Honk and colleagues studied the effect of exogenous T versus placebo in women who undertook the IGT. A women-only sample was chosen because the researchers claimed that knowing how much T would be needed to create an effect in men was impossible, implying greater knowledge of the dynamics of T in women. They used a double-blind, placebo-controlled crossover design, meaning that the women who initially received T before the gambling task were later retested after receiving a placebo, and vice versa. They found that T shifted women's decision-making pattern toward the disadvantageous strategy. Van Honk and colleagues interpret this as entirely in keeping with the larger literature on T and dominance, saying T is associated with high-risk, aggressive, and antisocial behaviors that ultimately yield greater resources. But if we follow the logic we just outlined, the findings instead suggest that these T-driven behaviors would lead people to squander their resources, accruing to reproductive disadvantage in the long run.[14]

The IGT is one of the very few measures that have been used in more than one study of risk-taking and T, making this an especially informative node in the research. Surprisingly, all of them repeat the contradiction that we just described. The evolutionary story once again covers the gap in logic, as van Honk and colleagues write: "Effects of testosterone

115

depend on situation or environment, thus in search for high status the hormone in our capitalistic society might have adaptively turned to materialism and greed. Indeed, proxies of current and prenatal testosterone correlate with the financial successes of traders on the stock market, which is argued to indicate that the hormone adjusts behavior in humans in instrumental ways to maximize personal profits."[15] They say this even though in research using the IGT, T is mostly associated with a "disadvantageous pattern of decision-making" that led to loss of personal "wealth" rather than gains.[16]

If you're not yet scratching your head over how studies using the IGT supposedly support the theory that T is the proximate mechanism carrying through the sexually selected pattern of greater risk-taking in men, contemplate this: women typically show "riskier" patterns of card selection on the IGT. It is perhaps so ingrained that risk-taking is masculine that finding a link between T and any measure of risk makes it irresistible to slot that finding into the grand evolutionary tale, even when that means forcing several square pegs into round holes.

Other People's Money

Despite these troublesome details, researchers remain convinced that the IGT is a useful tool for predicting risk behavior in the real world, especially where financial decision-making is concerned. According to psychologist Steven Stanton, an important wave of research emerged out of IGT studies to shine a light on a new sort of gambler: "professional investors [who] take significant risks in spite of large potential losses."[17]

The promising research to which Stanton was referring had been published by the neuroscientists John Coates and Joe Herbert in 2008 just preceding the worst US and European financial crisis in seventy-five years. The bankruptcy of Lehman Brothers, a global banking powerhouse, in September of that year prompted huge taxpayer-financed bailouts in the United States and elsewhere as stock markets tumbled worldwide: on September 29, $1.2 trillion in market value was obliterated in the biggest-ever single-day loss to that point. Panicking, countries closed their markets temporarily. Most commentators pointed to structural causes for the crisis, such as the widespread failure of regulators to rein in Wall Street corruption, high-risk financial products, excessive borrowing, and undisclosed conflicts of interest.[18]

But some pinpointed another possible cause: traders' biochemistry. In early 2009, at the World Economic Forum in Davos, Switzerland, a moderator on one of the panels posed a peculiar question: Would the world be in this financial mess if Lehman Brothers had been Lehman Sisters? The consensus among the women panelists, according to a *New York Times* reporter, was that if women ruled the trading universe, they "would have saved the world from the corrosive gambling culture that dominated many a trading room." Grameen Bank founder Muhammad Yunus, who won the Nobel Peace Prize for microcredit initiatives in Bangladesh and beyond, lending money largely to poor, rural women, agreed: "Women are more cautious," he said. "They wouldn't have taken the enormous types of risks that brought the system down." A European commissioner was "'absolutely convinced' that testosterone was one of the reasons the financial system had been brought to its knees."[19]

This idea didn't come out of nowhere. A year earlier John Coates, a former Wall Street trader-turned-neuroscientist, and his colleague Joe Herbert had published a study in the highly regarded *Proceedings of the National Academy of Sciences of the United States of America*. "Little is known about the role of the endocrine system in financial risk taking," the paper begins. The team sought to fill this gap by studying male traders in London, hypothesizing that on days when a trader had earnings greater than his average daily profit, his T would rise, and this would cause him to take greater risks. In one of the most extensively cited studies in the risk-taking literature, they claimed to find just that.[20]

Later, Coates expounded on a "universal biology of risk-taking" in his acclaimed 2012 book, *The Hour between Dog and Wolf: Risk Taking, Gut Feelings and the Biology of Boom and Bust*. The theory is based on studies of animal species as diverse as dogs and flies, in which scientists have shown that animals who won a fight or competition for turf were more likely to win the next one, a phenomenon called the "winners' effect." Coates writes, "Life for the winner is more glorious. It enters the next round of competition with already elevated testosterone levels, and this androgenic priming gives it an edge that increases its chances of winning yet again. Through this process an animal can be drawn into a positive-feedback loop, in which victory leads to raised testosterone levels which in turn leads to further victory." This result can be seen as positive, to be sure, but the confidence, euphoria, and greater tolerance for risk-taking can also lead to overconfidence and too much risk-taking. The "hour" in Coates's book's title refers to what he characterizes as a Jekyll-and-Hyde transformation: traders "became cocky and irrationally risk-seeking when on a winning streak, tentative and risk-averse when cowering from losses." T is "the hormone of economic booms"; cortisol is the "the hormone of economic busts."[21]

For the *PNAS* study, Coates and Herbert gathered a group of seventeen male traders, aged eighteen to thirty-eight, working at a single London firm. To maximize their chances of finding a relationship between hormones and trading behavior, they timed the study around the release of key economic reports, saying they hoped to capture "the times of greatest volatility" in the market. The team sampled the traders' saliva at 11 A.M. and 4 P.M. on eight days at times that book-ended trading activity and reported that traders' T was higher on days when they made more than their daily average. They also said that higher levels of T in the morning correlated with higher profits in the afternoon, while higher cortisol correlated with greater losses.[22]

The findings, they said, successfully tracked their hypotheses about T and cortisol—though it turns out that this required a significant retrofit. Their data didn't actually support the prediction that rising profits would lead to rising T. In fact, they don't even report running any analyses that would have addressed that hypothesis. This is the first clue that this study has been *p*-hacked.

Consider what they should have looked at given the hypothesis that T rises with increasing profits. To show this, they would first need a measure of T change from the morning to the afternoon, by simply subtracting the morning level from the afternoon level. Then they should have looked at those daily changes in T to see if they were associated with traders' daily profits. But they never report computing data for change in T over the course of a day. It's possible that they computed those data and saw nothing of interest—in other words, that the change in T wasn't associated with a change in profits. If that was indeed the case, that should have been the end of the study.

Instead, they averaged each individual's afternoon and morning T values for each day. Then they looked to see if that daily average T was associated with that individual's daily profits. While they found a statistical correlation, the comparison is simply wrong. Let's give them the benefit of the doubt that even though it wasn't what they said they were looking for, they had found an interesting link between daily profits and daily average T. As they note, that doesn't tell you whether T influences profits or profits influence T. To examine the direction of the relationship, they split every trader's set of days into two groups: one for when the man's morning T level was above his daily average, and another for when his morning T level was below the daily average. They found that profits were higher on days on which morning T was high compared to days on which morning T was low.

Once again, they've switched the subject. Recall that their initial hypothesis was that higher profits lead to higher T. If they were going to categorize days as "high T" and "low T" based on a single time point, they should have used the afternoon T value, which followed the profit-making, not the morning value.

By this point in the paper, it's clear that Coates and Herbert had abandoned their hypothesis. When nothing turned up in their search for correlations between daily losses and cortisol values, they tortured the data until they talked. After three strikes with their planned comparisons, they turned to other measures that might show a link between cortisol and losses. This part of the paper reads a bit like a detective story, as the researchers detail their meticulous search for correlations, signaling the exploratory nature of their process with phrases like "we therefore looked" and "we suspected that" and "consequently, we looked to see whether . . ." You can almost hear the ship creaking as they turn their analysis to meet their data. This is the very definition of *p*-hacking.

But the most stunning thing about the paper is the discussion, where Coates and Herbert manage to package their analysis and findings as if it were all seamless. Despite never even looking at whether T rises after profits do, their conclusions hark back to the winners' effect. They collected the temporal data that would have allowed them to test the hypothesis that higher profits one day would lead to higher T the next day, but instead they examined whether higher T one day was correlated with profits the next day (it wasn't). Nonetheless, they speculate that the correlations they found between high morning T and that day's profits could signal that higher profits lead to a long-term elevation of T, which in turn results in overconfidence. Voilà! The financial crisis has been explained.[23]

Despite its incoherence, Coates and Herbert's study is widely discussed as something that not only explains the great bust of 2008 but also gives fundamental insights about the embodiment of risk-taking behavior that links us to other animals and universal trans-species masculinity. "Risk is more than an intellectual puzzle," Coates wrote in a 2014 essay for the *New York Times,* "The Biology of Risk": "It is a profoundly physical experience, and it involves your body. Risk by its very nature threatens to hurt you, so when confronted by it your body and brain, under the influence of the stress response, unite as a single functioning unit. This occurs in athletes and soldiers, and it occurs as well in traders and people investing from home. The state of your body predicts your appetite for financial risk just as it predicts an athlete's performance." In an interview, Coates drew a straight line between finance and the way males of other species go from masters of their territory to out-of-control predators: "Animals go out in the open, pick too many fights, patrol areas that are too large and there are increased rates of predation. Risk-taking becomes risky behavior. That's exactly what is going on in Wall Street." T is the magical substance that keeps these natural links strong.[24]

Coates and colleagues provide some of the clearest examples of how the highly specific worldview of people who can invest disposable income for profit gets generalized as "human nature" via studies on T and risk-taking. In a 2010 article, Coates and his coauthors begin by explaining that studies using the challenge hypothesis and other insights from animal studies of hormones and behavior have had "questionable success." The researchers say that this is because, first, higher cognitive functions in humans "refract the effects of testosterone," and second, "the dependent variables in these studies, such as aggression, dominance, or status seeking, often cannot be defined or measured in humans with any objectivity." This is where studies of market activity come into play. The promise of such studies, they say, is twofold. First, "financial variables, such as profit, variance of returns, volatility of the market, can be defined objectively and measured precisely." Second, the best human analogue for the competitive and aggressive behavior observed in animals may be "competitive economic behaviour. Through its known effects on dopamine transmission in the nucleus accumbens, testosterone may well have its most powerful effects in humans by shifting their utility functions, state of confidence or financial risk preferences." As they see it, financial behavior is the sine qua non of risk-taking in humans, and conveniently, there are objective markers with which to investigate it that are readily available in the real world.[25]

Coates and colleagues suggest that the measures available are inherently objective because they are quantifiable, because numbers are involved. But this conveys a clarity that isn't there. As a researcher, one must sort through the nearly infinite array of financial metrics and choose the ones that best capture risk. Even then it still is not a straightforward matter of merely logging that number. Take, for example, the suggestion that a dollar amount is a simple and objective reflection of the degree of risk in a trade. The trades in Coates and Herbert's study ranged from £100,000 to £500,000,000, so it seems easy to put these on a scale and call the money at stake the risk. But who bears the risk? Is it the trader? The fund he is working for? The individual investors whose money the fund manages? The stakes may be relatively large, but the risk is always apportioned. Coates and Herbert gathered their data during the free-for-all that immediately preceded the global financial crash. In the wake of those frenzied trades, people all over the world lost their homes, life savings, and livelihoods. For the most part, traders did not personally bear the same burden, even though they were the ones taking the risks.[26]

Jens Zinn, a sociologist who studies the phenomenology of risk, argues that "it does not make sense to speak about risk-taking when the decision-maker is not affected by the outcomes (which instead affect others). This could be called risk-making (for others) rather than risk-taking and follows a different logic." The traders weren't monetarily unaffected, but the bulk of the money at stake was not theirs. In other areas of life, people whose risk-taking ripples outward to create negative consequences for others, or whose risks break social and legal rules, are sometimes labeled as "antisocial," "externalizers," or even sociopaths. These terms are used in some of the studies on T and risk-taking, but tellingly, the studies of risk-taking among traders, CEOs, entrepreneurs, and business students don't frame negative or irresponsible behaviors this way, even decidedly illegal activities like insider trading, options backdating and tax evasion, or cheating in business negotiations.[27]

The idea that trading and other financial behavior is the best measure of human risk-taking reflects an idiosyncratic worldview, to say the least. The Swedish study of investment behavior provided the commonsense insight that the risks people take depend on the resources they can command in the first place. Let's recap what we've learned so far. Two prominent nodes of research, studies linking T and occupation and studies using the Iowa Gambling Task, both directly contradict what is supposed to be the underlying evolutionary theory that ties risk-taking to T. The blockbuster study by Coates and Herbert was p-hacked and is logically incoherent. And within the domain of economic risk-taking, which is supposed to be a very important and well-defined subset of the risk-taking research, the concept of risk varies within and among important studies, even those that cite each other as mutual support.

The Big Messy Hairball

Annie Taylor fits a common idea of what risk is—zooming on a motor-cycle, betting one's life savings, jumping from a plane, sending oneself over a waterfall in a barrel. By any account, her plunge involved a huge phys-ical risk, and it might appear that it was among the riskiest things a person could do. But Taylor was an aging widow, and poverty presented its own very real risks to her. Risk doesn't inhere in any particular behavior, but rather is expressed by the full circumstances and context in which actions and reactions take place.

The narrative of risk-taking and T in the studies described earlier in this chapter makes finance look brave and bold, the pinnacle of risk, and naturally masculine, while obfuscating the real structure of wealth and business. Cordelia Fine has joked about the inherent sexism in the busi-ness lingo of taking "big, hairy, audacious risks," so perhaps it's appro-priate that the concept of risk in this literature looks like a big messy hair-ball. We've shown how messy the construct is even within the narrow domain of financial risk-taking. And when you move outward to include other kinds of behaviors that get lumped in the category of "risky," it's no longer clear what common thread holds these behaviors together. Fine cites Cass Sunstein's observation that risk-taking is an "unruly amalgam of things" that includes "aspirations, tastes, physical states, responses to existing rules and norms, values, judgments, emotions, drives, beliefs, whims." Under the rubric of "risk," studies we examined look, for ex-ample, at sexual behaviors but also sexual attitudes; taking physical risks but also having the willingness to take them; and, in one especially eclectic study, fifty items that ranged from "ethical risks" like shoplifting to "rec-reational risks" like "engaging in dangerous sports," "financial risks" such as "investing 10% of your annual income in a very speculative stock," "social risks" like "speaking your mind about an unpopular issue at a social occasion," and "health risks" including "eating expired food prod-ucts that still look okay." It's a lot to crowd together under the single umbrella of "risk."[28]

Some researchers focus on "sensation-seeking," a personality trait that supposedly underlies "risky behaviors," using Zuckerman's Sensation-Seeking Scale (SSS-V). The SSS-V consists of four subscales: Thrill and Adventure Seeking, Disinhibition, Experience Seeking, and Boredom Sus-ceptibility. The subscales sound logical, but the examples reveal a cultur-ally specific hodgepodge, ranging from some of the usual suspects like scuba diving and water skiing to esoterica such as liking "earthy body smells," "'sexy' scenes in movies," and "the 'clashing colors' and irregular forms of modern paintings" and social experiences such as "associating with flighty rich persons in the 'jet set'" and "meet[ing] some persons who are homosexual." Aside from being amusingly dated, the items on the SSS-V suggest that certain activities and material things are in and of themselves thrilling or boring. Picasso-lovers may have been risk-takers in the 1950s, but are they now? What of the poor Rembrandt-lovers? Are they to be forever seen as the cautious soft-shoed fuddy-duddies of the world? Associating with "flighty rich persons" may be a social risk in some socialist circles, but in other groups one can befriend rich people with no apparent cost to one's social standing. It's impossible to find the through line among these tastes and behaviors. And yet once a researcher ventured to hypothesize that SSS-V scores might be predicted by T, these behaviors all had the potential to be underwritten by that most powerful of hormones.[29]

One of the most widespread assumptions in the research is that risk-taking is fungible: whether it's dollars or social humiliation on the line, a risk-taker is a risk-taker. Such a view assumes that "appetite" or propensity for risk will manifest in every domain of life, whether health, finances, career, social life, ethics, or recreation (the last of which spans diverse activities like gambling, drinking, using drugs, sex, and driving, to name a few). To the contrary, Fine presents evidence showing both that risk-taking in one domain does not predict whether someone will take risks in another domain, and that the same person will take different risks in different contexts. Some researchers examining T's relationship to risk also recognize that there are multiple domains of risk, but even when they disaggregate risk into distinct domains, they retain the notion that "risk" is best captured in any domain by the more masculine choice, a problematic idea to which we return later.[30]

The risk concept can be complicated further by looking to researchers who aren't as interested in finding a common cause for risk as they are in understanding the phenomenon itself. The sociologist Jens Zinn, for example, "tentatively distinguishes" three general motivations related to risk-taking: "risk as an *end in itself,* as a *means to an end,* and as a *response to vulnerability.*" His "tentatively" signals that these categories aren't firm: the same behavior maybe be underwritten by multiple motivations, or it might not in practice be possible to distinguish how risk-taking as an end in itself is different from risk-taking as a means to an end. Zinn notes that taking or avoiding certain risks is never about detached cost-benefit calculations but rather is about such things as "building and protecting a meaningful identity, engaging in intimate relationships, securing a good income, making a valuable contribution to society, and building and continuing friendship relationships." Thus, to understand both the meaning-making that taking risks entails and the distributions of material stakes that are on the line requires a structural view of risk, not an individual psychological or psychobiological one. The highest levels of social institutions and processes shape risks because they create the landscape of options within which people operate and embed specific actions with meanings relevant to social statuses such as gender, race, class, religious and political affiliations, and more.[31]

Who "chooses" the extremely risky occupation of coal mining, for example? Being a coal miner is a source of pride for people who have familial and regional histories in the industry, but not everyone in those regions or families opts to go "down the mines": it's still a job for people with few if any other options for earning a living. Discursive structures also obscure the risk inherent to certain behaviors or occupations. The dangerous things that some groups of people do as a matter of course because of role expectations and material constraints are not typically examined as risks. For example, the epidemiologists Karen Messing and Jeanne Mager Stellman have documented the surprisingly extensive toxic exposures and high accident rates involved in domestic labor, a pattern obscured by the widespread notion of home as a "safe place."[32]

So it's not just desperation that makes people do dangerous things, it's expectations, norms, and the daily circumstances of inequality. Consider the mundane behavior of asking for directions. This isn't inherently risky, but the context can make it so, depending upon who you are, where you are, and whom you ask. In recent years, there have been innumerable accounts of black people going about their normal daily routines—asking directions, waiting in a cafe for business associates, playing in a park, driving, resting in a student lounge, and barbecuing are a few examples—who have faced extreme consequences ranging from arrest to assault to murder, at the hands of both civilians and the police, so many of whom have been white. Likewise, as Fine has pointed out, childbirth for a woman in the United States is about twenty times more likely to be fatal than skydiving. Even this surprising level of risk masks huge disparities by race and class. Black women continue to experience devastatingly high maternal mortality rates: more than four black women die for every thousand live births, while just over one white woman does. Native American and Alaskan Native women die at twice the rate of white women.[33]

Status, like wealth or poverty, shapes the actual risk involved in particular behaviors. Because status also affects T levels, as we've explained in earlier chapters, studies linking risk-taking with T should include status in their models, but generally don't. In 2017, sociologist Susan Fisk and colleagues published a broad review of the literature on T and economic risk, putting special focus on status. Noting that the evidence of a positive association between T and risk-taking was tempered by an "abundance of null results," they said that even this might overstate the relationship because researchers have used a simplistic model in which T would have a direct effect on risk-taking. But multiple lines of evidence suggest that status is an important mediating variable that might explain some or even all of the apparent association between T and risk-taking. In other words, "given that testosterone is a social hormone with a reciprocal relationship with social status, and social status has been found to drive risk-taking behavior," the positive relationship between T and risk-taking might be spurious.[34]

Fisk and colleagues found that "risk taking is less risky for those with high status." Status especially affects the way other people evaluate someone who fails in their risk-taking, with high status mitigating the failure. Lower-status people who fail are seen as being responsible for their failure, while higher-status people are seen as having bad luck. Examining race as an important aspect of status, Fisk and her coauthors observe that evidence is mixed regarding whether race affects the way people take economic risks, but race does affect how people perceive the risk involved in situations they face. White people, especially white men, "generally perceive the world as less risky than women and minorities, even when they face the same level of risk." It's a good idea to pause for a moment and note that there is already a contradiction here in the idea that risks can be held constant across gender and racial groupings, given that the researchers also note how lower-status people will suffer greater losses if the risks that they take fail. Moreover, the differential stakes in similar situations across status groups (including people of varying genders, races, classes, abilities, and more) surely go beyond how someone is judged when she succeeds or fails, encompassing, for example, social capital and deep economic supports that can cushion losses and/or amplify wins. People's social status positions them differently in relation to specific risks: higher-status people are exposed to losses with the disruption of the status quo, while lower-status people are given the possibility for gains. Even laboratory games that present apparently equal monetary stakes to participants can't equalize the relative importance of the risks they face: $50 might be negligible to some, but it represents a week's worth of groceries to another.[35]

Researchers' assumptions about what risk-taking is, and their measures of risk, suture risk-taking to masculinity in a way that virtually guarantees they will find greater propensities for risk in men. Thekla Morgenroth, Cordelia Fine, and colleagues recently showed that this gender bias in measures operates at the level of very specific risk scenarios, not at the level of "domains" such as physical, financial, or social risks. For example, typical items in the domain of financial risk involve gambling behavior, such as "betting a day's income at a high-stake poker game." These researchers developed new items that were similar in terms of stakes but less obviously gendered, such as "buying a flight from a less reliable airline that often cancels its flights but is 50% cheaper when flying to an important event (which she/he will miss if the flight is canceled)." They conclude that findings of "greater male risk-taking in a particular domain can't be considered to be generalizable to other forms of risk-taking, even within that domain of risk." Given the gender bias in item selection that Morgenroth and colleagues have demonstrated, it is inevitable that studies will, on balance, overestimate the association of risk-taking with male gender.[36]

Still, researchers who focus on T generally theorize rather simply that the higher level of exposure to T in males maps easily onto the apparently greater risk-taking in men. To the extent that gender socialization is considered, it is frequently seen as "masking" the true underlying differences that sex-specific biologies would dictate. Fisk and colleagues argue against such gender-essentialist interpretations, citing evidence that social phenomena influence the size and even existence of gender differences in risk-taking. For instance, differences in status have been found to fully account for apparent gender differences in risk-taking, and men's desire to appear masculine also encourages them to take greater risks. What's more, the typical model ignores evidence from reciprocal-effects studies that show T to be responsive to status instead of driving status. That is, high-status behavior increases T, including in women. Rather than a simple linear model that suggests T leads directly to risk-taking, we have sketched

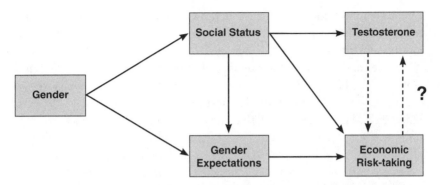

FIGURE 5.1 Possible model for relationship among gender, social status, economic risk-taking, and testosterone.

(Adapted for the authors by Sheila Goloborotko from Susan R. Fisk, Brennan J. Miller, and Jon Overton, "Why Social Status Matters for Understanding the Interrelationships between Testosterone, Economic Risk-Taking, and Gender," *Sociology Compass* 11 (2017): e12452, 1.)

out a more realistic model based on the work of Fisk and colleagues (Figure 5.1).[37]

Taking Biology Seriously

We've spent some time exploring the ways that studies of risk-taking and T give short shrift to sociopolitical variables, but it turns out that researchers often aren't careful about the biological variables in question, either—starting with testosterone.

Increasingly, investigators look at multiple measures of T in the same study, including T at different time points, proxies of prenatal T as well as circulating T, and T in relation to other hormones such as cortisol. In the best-case scenario, investigators make adjustments that get them closer to accurate and mechanistically precise descriptions of how T works. If great care is not taken, though, the risk is that proliferating the versions of T on the input side of causal equations creates more opportunities for *p*-hacking: there are simply too many options for what analyses can be run even under the rubric of a straightforward initial hypothesis such as "T increases risk-taking." Researchers often proceed as if each version of T in their study is just a different measure of the same abstract thing. But T's multiplicity means that each measure is pointing to a different phenomenon: a different measure can't be chosen simply because it is more convenient or because it is the one that yields statistically significant associations.

Coates and Herbert's famous study is a prime example of how T's multiplicity can bolster weak findings. Their hypotheses and measures involved morning T, afternoon T, and mean daily T for each of the eight study days. Then there's the T that their hypothesis implied but they didn't compute: change in T. But as we saw, the multiple versions of T in that study support a shell game that makes it difficult to keep track of exactly how any particular measure is related to their hypotheses. While they hypothesized that rising profits caused rises in T, they only used mean daily T and morning T—versions of T that could not reflect a response to the day's profits. There was no room in their hypotheses for either of those versions of T, so by introducing them, the team created more opportunities to find statistically significant associations. There are additional multiplicities to consider, including how T is collected and measured. Nine of their seventeen subjects had to chew gum to produce enough saliva for the hormone sampling. Quite a few researchers still use gum for this purpose, but there have been hints since 1989 that chewing gum affects the measurement of T, and recent research suggests that those effects are large.[38]

Another way that researchers examine multiple versions of T is to incorporate the organization-activation theory, which holds that T affects behavior first by prenatal organization of the brain for specific masculine behavioral and cognitive proclivities. Later, circulating T "activates" more masculine behavior patterns. Thus some researchers have looked for evidence that risk-taking is shaped by fetal T instead of or in addition to circulating T. No one has direct evidence of fetal T exposures, so researchers use an array of proxies, such as the 2D:4D finger length ratio, facial masculinity scores, and left- versus right-hand preference, all of which are somewhat controversial as indicators of prenatal T. Some researchers are also beginning to turn to genetic data, specifically gene sequences related to androgen receptor activity, as a proxy for variation in how efficiently different bodies use circulating T. All of these proxies require adjusting the hypothesis underlying a study to account for the particular timing and pathways through which T might exert effects on risk-taking. Or rather, they should. But multiple identities of T in studies on risk-taking are usually downplayed in favor of a seamless narrative that there is a lot of evidence linking T to risk, while neglecting the finer points that might high-light breaks and contradictions between studies.

Finger length ratio, called 2D:4D, is the most common proxy for fetal T that has been used in risk-taking studies. The rationale for its use is that testosterone during early development affects the relative length of digits in experiments with nonhuman animals, and men in most populations have a lower 2D:4D ratio than women, meaning that their index fingers are relatively short compared to ring fingers, on average. Following the organization-activation hypothesis, any measure of prenatal T should modify relationships between circulating T and risk-taking. But investigators don't hew closely to this hypothesis, instead incorporating digit ratios as just another chance to find a positive association between T and risk-taking. Even with the flexibility of the hypothesis afforded by multiple measures of T, the evidence regarding digit ratio and risk-taking is, on balance, in contradiction to the expected hypothesis: a recent large meta-analysis found that a more "feminine" finger ratio was associated with higher risk propensity. But the theory has enormous appeal. One of the most frequently cited studies on risk-taking and T is a 2009 study by psychologists at the University of Chicago and Northwestern University investigating both adult circulating T and markers of prenatal T in men and women business students. Economist Paola Sapienza and colleagues reported that T levels during both time frames shape risk-taking, proclaiming right in the title of their paper that "gender differences in financial risk aversion and career choices are affected by testosterone." Another team of psychologists, Daphna Joel and Ricardo Tarrasch of Tel Aviv University, issued a brief but devastating critique shortly after the original study appeared. Joel and Tarrasch point out that no correlational study could support the causal claims that Sapienza and colleagues make, and also call out inappropriate statistical tests and other problems with the study. But the rebuttal has been almost totally overlooked. While the original study has amassed 274 citations, the rebuttal has but four, even though both were published in the same prestigious journal.[39]

Some other risk-taking studies look at a fundamentally relational T, one that is dynamically associated with other hormones, typically cortisol (C) but sometimes estradiol. We haven't done a systematic analysis of how researchers frame the relationship between T and C in their hypotheses about risk-taking, and how that framing tracks onto measures and onto interpretation of findings. But a quick sampling of studies that discuss how

T and C might work together gives a taste for the variety of hypotheses and findings in this subset of studies. One popular way to include C is via the dual hormone hypothesis. To put it in the simplest terms, this hypothesis suggests that T and C will have inverse relationships to risk. A variation on this is that C, thought of as a "stress hormone," is related to anxiety and inhibition, and therefore is thought to attenuate the effect of T on risk-taking. In other words, T will be related to risk-taking only when C is low but not when C is high. Generally, T is the hormone seen as driving risk, while cortisol acts as a moderator; sometimes this is reversed and researchers suggest that C drives "risky behavior" (through "cortisol reactivity") while T moderates that effect. Recently, a new "coupling" hypothesis has emerged, which suggests that, depending on the context, T and C may rise together and even mutually activate each other, or conversely may mutually inhibit each other.[40]

Once again, there's an inherent tension between the desirability of moving toward a more sophisticated and complex model, on one hand, and introducing so many variables so that there are too many chances to find positive associations, and interpretation becomes murky. The quick scan above shows that there's a model available for nearly every pattern of findings, so everything seems to support a workable theory. But the predictions about how the two hormones should relate to risk are all over the map. Thus, without being comprehensive, there's a strong suggestion here that C is often introduced opportunistically, again simply adding to the variables that might show correlations and keep alive the idea that hormones drive risk-taking.

Coates and Herbert again provide a prime example for how adding cortisol to the model can provide cover for p-hacking, that is, for multiplying opportunities for finding statistically significant correlations. Underneath a very elaborate discussion of markets and how they relate to profits, losses, and traders' stress, there's a striking disconnect in how T and C are analyzed. Early in the study they argue that most of their study participants only care about US markets (who report big announcements around 1:30 P.M. in London). Theoretically, all personal and market events before that time would just be random, independent of market activity. On the other hand, they conclude that market activity predicts cortisol variation over the day. To make this argument, they use German Bund market activity and correlate that with variations in cortisol across the day. While the theory is supposed to be about T and C acting and reacting in concert in a coordinated context, there's a bait and switch: the data show T and C operating independently in disconnected contexts.

Taking biology seriously means holding researchers accountable for using hormone measures that track their initial hypotheses, and it also means that when data don't support the hypothesis, investigators must be clear about the rationale for moving to another hypothesis or model. Small and seemingly insignificant shifts in wording signal theoretically important differences in how researchers understand the biology that underwrites risk-taking. For example, if the hypothesis posits that T is modified by cortisol but the discussion indicates that T is a modifier of cortisol's actions, then there has been an important shift in the underlying theory. We find a lot of this sort of slippage, and it shifts the literature to which findings must be related.

One lesson we hope to have conveyed is that investigators who study T in the context of risk-taking step into a mass of hypotheses that can be hard to track. It's not just that risk-taking is a complex and, in the T literature, poorly specified construct. T is also complex and multiple. Treating both risk and T as if they are simple covers gaps and contradictions in the studies, enabling global statements about T and risk. But studies of T and risk-taking don't hold together: many of them contradict the evolutionary tale they are supposedly based upon; some of the most influential studies have astonishing mismatch between their hypotheses and their data; and even with considerable evidence of p-hacking, the overall evidence for T affecting risk-taking is "erratic" at best.

• • •

STUDIES ON T AND risk-taking collectively contribute to the shopworn narrative that T supports social hierarchies: men's T-driven boldness is what drives them to the top. It's not just gender hierarchy that gets naturalized: people in finance and successful entrepreneurs are somehow biologically bolder and therefore naturally fill those positions of economic power. Across these studies, T is both flexible and deterministic—depending on who you are, T works differently to cement multiple hierarchies and exclusions. When researchers explore those risks seen as necessary for the status and "resource accumulation" that increase reproductive success, they study people who are already at or near the top of social hierarchies: business students, financial traders, and entrepreneurs. When they explore "risky behavior" or "antisocial" risks seen as fundamentally and inherently negative, they study adolescents and children, prisoners, and low-rank army veterans. These latter samples are more ethnically and economically diverse than the samples in financial risk studies, where the subjects are mostly white.

Why do the narratives matter so much? We need frameworks that will relate variables and mechanisms to something we already understand. In other words, science needs stories. But do stories need science? Scientism encourages us to seed narratives that are far afield from research with science-y details that make them seem more alive, plausible, and engaging. Stubbornly picking at the constructions of risk and T in these studies, we've taken on the role of party poopers, interrupting an elegant and apparently appealing narrative that powers a lot of research and offers answers to several important social questions: Why did the financial markets collapse in 2008? Why are there so few women traders? Why do adolescents seem to take crazy risks?

Science is not only storytelling, and this is why we insist on following the constructs and thinking about which specific version of risk or T is being mobilized in particular studies. But one story—the idea that T, via sexual selection, has ensured the pairing of maleness with a whole suite of traits and behaviors—is the glue that holds a whole body of research together. Narratives and data can also be fit into the rubric of "floaters" and "sinkers": floaters are the stories and bits of data that get picked up from researchers' discussions, abstracts, and titles and get cited in subsequent research, and the sinkers are the bits that don't fit, the awkward gaps between hypothesis and data, the multiple analyses that are done offstage and never again mentioned. The data on risk-taking and T are weak at best, and certainly chaotic. But the narrative has a pleasing parsimony: T increases risk-taking.

6

PARENTING

IN 2011, A STUDY by the biocultural anthropologists Lee Gettler, Christopher Kuzawa, and colleagues demonstrated for the first time that T dropped after men became fathers. Earlier studies had found that married men, and especially fathers, have lower T than their single and childless counterparts, but those studies were cross-sectional snapshots of a population at one point in time, and thus unable to address causation. Maybe the link observed between lower T and fatherhood was an artifact of aging: T declines with age, while the likelihood that a man is also a father increases with age. Or maybe men with lower T are more likely to enter stable partnerships or become parents in the first place. Gettler and Kuzawa collected data from men before and after they became parents, which decisively settled the debate: fatherhood was driving down T.[1]

The media and the blogosphere buzzed with anxiety. What did this study say about the manhood of fathers? About nature's plan for men? About whether it was good or bad, healthy or emasculating, for men to change the baby's diapers? Some headlines cheered the study, proclaiming that men are "biologically wired to care for children" and that fathers' drop in T both is "for the good of the family" and "may protect men from chronic diseases" that have been linked to high T, such as prostate cancer and high cholesterol. Other headlines, though, lamented that "being a dad makes less of a man," and seemed to confirm dads' darkest fears by asserting that the "plummet" in T levels explains "why dads' sex drive is stuck in reverse." The article about fathers' sex drive, which was reprinted in multiple papers, revived the ubiquitous, though erroneous, conflation of libido with T levels. The study, wrote the reporter, "found that newly partnered men who did not become fathers had similar sex drive levels to single men. But partnered fathers showed an average 34% decline in tes- tosterone." In fact, the study didn't report anything related to "sex drive," either in terms of sexual interest or in terms of activity. Nevertheless, the study had clearly struck a nerve, and the widespread association of T and libido made the reporter's claim seem obvious, even though it was wrong.[2]

Peter Ellison, a Harvard biologist whose work on T and relationships helped to set the stage for the study by Gettler, Kuzawa, and their team, had anticipated a backlash and tried to cut it off at the pass. Interviewed for a front-page story in the *New York Times,* Ellison said the "real take-home message" is that "male parental care is important . . . enough that it's actually shaped the physiology of men." He also speculated that Amer- ican men, in particular, wouldn't be happy to hear about this dip in T: "'American males have been brainwashed' to believe lower testosterone means that 'maybe you're a wimp, that it's because you're not really a man.'"[3]

As if on cue, a few days later a second story in the *Times* gave barely tongue-in-cheek voice to the fears of emasculation that Ellison had pre- dicted. The journalist Alex Williams reported that the study had been im- mediately dissected in numerous fatherhood blogs and had become a topic of conversation in bars and dads' groups. One of the men Williams interviewed was Robert Fahey, a father of two from Burlington, Massa- chusetts, who was not happy: "A study like this implies you are scientifi- cally less manly just when you'd like to think you've hit a new plateau of manhood. . . . You've spread your seed, so to speak, and joined the ranks of your own father." Now, he fretted, "not only are you a dork when you lapse into goo-goo talk, but now you're less of a man scientifically."[4]

The study that set off this firestorm had begun years earlier in Cebu City, the Philippines. Gettler, Kuzawa, and colleagues piggybacked on a large health study to track T levels in young men as they transitioned into adult partnerships and parenthood. In 2009, their team showed that young fathers in committed relationships had lower T than both single and "pair bonded" non-fathers, building on a number of other studies that had al- ready shown that fathers and married men usually had lower T than non- fathers and single men. Given the long-standing assumption that T makes men less interested in stable sexual partnerships and more interested in sexual variety, many researchers suspected that T was most likely driving the relationship seen in the study data. But Gettler and Kuzawa found that men who began their study with higher T were more likely to be in committed relationships and have children four years later than their lower-T counterparts. Further, being involved in direct caregiving seemed to reduce T even more: those fathers who reported being a primary caregiver for their children had the lowest T. Because the greatest differences in T levels between the most-engaged fathers and others was observed in the evening, after a day of caregiving, the causal direction seemed even more plausible.[5]

How could the results of a study done halfway around the world deal such a blow to Fahey's sense of manhood? From straight news reports to satires, coverage suggested that the study spoke to a widespread fear that dads who do hands-on childcare aren't really men anymore. Alex Williams might have been aiming at hyperbole, but he nevertheless captured the general zeitgeist when he wrote, "In a Mr. Mom era, where society encourages (and family schedules often demand) that men enthusiastically embrace a 50–50 split of every parenting duty short of breast-feeding, the question on many fathers' minds is whether all of their efforts to be the ideal contemporary man are also making them less of one."[6]

If you read Lee Gettler's theoretical articles on T and parenting, not to mention his own blog posts and interviews after this study came out, you can almost hear the sound of him ripping out his hair. Far from showing that fatherhood makes men less manly, Gettler thinks, the studies show that human males have been involved with caring for their babies since early in our species' history. "If this weren't something that had been normative in humans for the last 100,000 or more years, there would be no reason to expect this decline in testosterone," says Gettler. One conclusion, delivered in separate interviews by both Kuzawa and Gettler, is that men are "'biologically wired' to help raise children.'" "This is important," Gettler added, "because traditional models of human evolution have portrayed women as the gatherers that take care of the kids and stay behind."[7]

•　　•　　•

Most researchers have framed the role T plays in relationships, both sexual/romantic and parental, in terms of "investments." According to the various theories, T brokers the way that people—in most models, men—divide their time and resources. More generally, the idea is that organisms, including humans, have only so much time and energy to allocate to the various tasks of growth, maintenance, and reproduction, and any energy they invest in one of these basic functions will necessarily decrease the energy they can allocate to the others. In evolutionary terms, it's called a "life history trade-off."

As with aggression research, most researchers studying T and "pair bonds" or parenting draw on the challenge hypothesis, a theory developed to explain seasonal variation in T levels among birds. Researchers posit that higher T in the "mating" stage enables men to compete for mates and reproduce, while lower T in the parenting and partnering stage enables men to be engaged and nurturing, helping to ensure that offspring survive and thrive. A second model connects T to reproductive investment strategies by drawing on "r/K selection theory," an evolutionary theory about reproductive strategies that prioritize a high number of offspring versus those that emphasize a high level of investment in existing offspring. Recently, psychologist Sari van Anders and colleagues offered a third model that built on the challenge hypothesis but interprets T's role vis-à-vis specific social bonds instead of in terms of facilitating investment trade-offs; van Anders further extended the hypothesis by insisting that a robust model must explain T's behavior in women as well as men. Across all three models, researchers use data on T and parents to rethink important underlying theories about sex/gender, T, and evolution.

In what follows, we explore the meanings of "facts" about lower T in people who are parents and those who are in stable sexual or romantic relationships—groups that obviously overlap—by exploring how these facts are used, and contrasting the approaches and effects of the three basic models. The data get taken up in profoundly different ways, demonstrating that emerging scientific facts about T have flexible meanings and contribute to widely divergent discourses about the nature of gender and racial difference.

Most of the earlier chapters examine familiar facts about T to see how they were produced, and we do less of that in this chapter. In part, this is because lower T among those who are parents and pairbonded isn't yet deeply entrenched in T's standard biography—this is still emerging science, and much of it seems to go against the grain, at least judging from the popular reception of studies. We do less poking and prodding at people's measures and statistical practices, but that doesn't mean that research practices in this subset of studies should be taken at face value. Instead, it indicates our slightly different aim in this chapter: showing how

current evidence about the relationship of T to social bonds and investments in parenting can simultaneously support the old yarn and take T's
story in new directions.

Revising Evolutionary Theory

Gettler was still a graduate student when he published an ambitious piece
on T, evolution, and fatherhood in the American Anthropological Association's flagship journal, *American Anthropologist*. Titled "Direct Male
Care and Hominin Evolution: Why Male–Child Interaction Is More Than
a Nice Social Idea," Gettler's article proposed a new model of human evolution in which direct male care of offspring was pivotal.[8]

Gettler's reworking of the early parenting tale is especially indebted to
the anthropologist Sarah Blaffer Hrdy's arguments about alloparents,
meaning individuals who provide direct care for children other than their
direct descendants. Hrdy assembled diverse lines of evidence ranging from
paleoanthropology to contemporary primate studies to cultural anthropology to convincingly argue that distribution of child-rearing tasks to
alloparents was perhaps *the* pivotal innovation in human behavioral evolution, giving rise to a uniquely cooperative species. Gettler's contribution was to use the concept of alloparents to rethink the role of human
fathers.[9]

Traditional evolutionary models of men's parenting have focused on
three simple components: sperm, physical protection, and "provisioning,"
or providing men's offspring and their mothers with calorie-dense food
obtained by hunting. Observing that "direct male care [of children] is
common in many human populations," Gettler set out to show that if this
pattern emerged in early human history, it could explain one of the long-
standing puzzles of human evolution: why are humans as fertile as they
are, given how "costly" human babies and children are to raise? Cost, in
this case, is expressed in terms of calories, both enough to keep the baby
alive and enough so that caring for the baby, whether that means breast-
feeding or carrying the little darling, doesn't kill the caretaker.

Moreover, the relative energy costs of reproducing would have increased, rather than decreased, over human history. As early humans
evolved to be larger and developed relatively bigger brains, their metabolic rates would have also increased, resulting in the need for still more
calories. With these changes, pregnancy and lactation would have become

even more "expensive" from an energy standpoint. How could such an energy-hungry species survive and thrive? As indicated by differences between humans and our nearest relatives, key related adaptations that seem to have taken place are that human young are weaned earlier, and subsequent births come with greater frequency compared to other primates. The short interbirth interval in modern humans is especially notable because typically interbirth intervals are long in bigger primates. Humans, while relatively large, have an interbirth interval that is nearly 50 percent shorter (about 1,114 days) than our closest relative, the (smaller) chimpanzee (about 2,013 days). How did human females manage to wean their young so early, when the young remain dependent for so long?[10]

The answer, according to many scholars, is in what fathers bring to the table—literally. In the economic language common in evolutionary theory, this is known as "male investment." Gettler agrees that male investment is central, but thinks most scholars have been blinded by traditional thinking about what paternal investment looks like. The outsized idea of "dad the provider" crowds out evidence that human dads are often direct caregivers; Gettler maintains that other theorists have been fixed on the idea of male provisioning because they've been too dedicated to the idea that a sex-based division of labor was central to how early humans survived.

With this revisionary tale, Gettler joins Hrdy and others in taking on some giants of evolutionary theory, including Owen Lovejoy. Renowned for his work reconstructing our most famous human ancestor, Lucy, Lovejoy is one of the most important theorists to give an explicit rationale for how male provisioning would have worked to support early human evolution. In Lovejoy's model, the sexual division of labor was a solution to several problems humans faced at roughly the same time in our development. Besides increasing the metabolic costs of pregnancy and lactation, larger brains meant that human babies were born at an earlier stage of development, meaning that they were vulnerable and dependent for longer. Meanwhile, changes in the environment meant that humans had to forage over a greater area to get enough food. By Lovejoy's account, this resulted in sex-specific "foraging niches," with males covering a much larger territory relative to females. In these cooperative, monogamous pairs of males and females, the females would specialize in intense parenting behavior, but the males couldn't do that without sacrificing their greater foraging range. The "simplest solution," he pronounced, was an arrangement where females would provide direct care for the young, and males would bring home the bacon. The upstart Gettler drily commented that this "simplest solution," though, "runs counter to observations of paternal investment in many human cultures and nonhuman primate species."[11]

Lovejoy's model of different foraging niches is essentially the story of "man the hunter, woman the gatherer"—a staple of evolutionary psychology. But not only does this rendition contradict contemporary cross-cultural evidence that fathers in many cultures provide direct care, it also doesn't fit current data on how and when hunting evolved. Big-game hunting is a fairly recent undertaking, a fact that has been overlooked because of a classic error in traditional evolutionary thinking. Many anthropologists and evolutionary psychologists have looked to modern-day foraging people and other so-called simple societies as if they can stand in for or embody the lives of early humans. Using observations of contemporary forager societies, they have tried to deduce how the additional calories available from men's hunting would enable women to focus on child-rearing and reduce the spacing of births. However, drawing especially on research by the anthropologist Frank Marlowe, Gettler points out that modern-day hunter-gatherers use technology that is far superior to anything available to early humans. The sort of big-game hunting on which theories of the sex-based division of labor are built requires tools that weren't developed until far later in human history—perhaps as recently as 80,000 years ago, and certainly not any earlier than 400,000 years ago. That's recent in evolutionary terms.[12]

In taking on the idea that provisioning is the main way that human fathers take care of their offspring, Gettler and company are doing more than making it seem "natural" that dads should share the childcare load. They are shaking up ideas about when and how a sexual division of labor entered our species. For example, Gettler sides with Frank Marlowe and others in believing that early human foraging was not a sex-specific affair. Instead, they posit that cooperative foraging encouraged the development of (heterosexual) pair bonds first, and a division of labor followed. The "man the hunter, woman the gatherer" model they are challenging is shorthand for sex-specific core survival behaviors that supposedly spawned an entire suite of sex dimorphisms in human psychology and behavior. And that's where this story intersects with T.

In this instance, Gettler is using data on how T behaves in contemporary men to revise a fundamental part of the evolutionary narrative that provides the backdrop for understanding T as a mediator of behavior. Usually, the evolutionary narrative is bracketed and taken for granted as the already-settled part of this scientific web of hypotheses and data. Gettler couldn't have made this bold move if there hadn't already been a strong cohort of researcher/theorists pushing hard against the settled "dad the provider/mom the nurturer" story and its close corollary, "man the hunter/woman the gatherer." Gettler overlaid data on T onto an existing debate in which other data and hypotheses, such as Marlowe's reinterpretation of evidence about early hunting and Hrdy's concept of allomothers had already opened up the possibility for rethinking the history of human sexual division of labor. When did this division take hold in our lineage? What is the precise content of the sex-specific labor for females and males, respectively? How strict is the division, and to what extent is that division a universal across human groups? These questions don't necessitate a wholesale rethinking of gender roles and evolution, but the answers affect how we think about gender, emphasizing that our evolved heritage involves flexible and overlapping gender roles rather than deep and strict divisions.

Just as a long history of involved fathers emerges from this story, so does the new figure of "woman the hunter." Not a spear-wielding woman warrior, exactly, but someone who had to develop the visual and spatial skills to track and kill small- and midsized game, as opposed to someone whose skills were about identifying and gathering stationary foodstuffs and having a monopoly on emotional and physical nurturing of other humans.

As but one example of how this shift in evolutionary narrative redounds upon contemporary stories about T, consider David Epstein's popular book *The Sports Gene*. In a chapter exploring sex differences and athleticism, Epstein points to T as the proximate mechanism that carries ancient sex-specific adaptations forward into modern competition. Recounting the stock tale of a sex division of labor in early humans that had men hunting big game while women were "tuber hunters," Epstein asks with a presumably straight face, "Why are women athletic at all?" He continues, "Like our male forebears, our female ancestors needed to be athletic enough to walk long distances, carry kids and firewood, chop down trees, and dig up tubers. But women were far less likely to fight, to run, or to push the capacity of their upper body strength with strenuous activities like tree climbing." Reading Hrdy gives the quite different impression that women protecting themselves and their infants not only ran and climbed trees but frequently did so with "expensive hitchhikers"—

babies—on their backs. Still, Epstein answers his rhetorical question with the suggestion that women's athleticism is not the result of adaptations that increased their own fitness but a by-product of the fact that male fitness depends so centrally on physical prowess and a competitive, focused nature. Just as our shared ontogeny has left men with vestigial nipples, it has left women with vestigial athleticism. Beginning with man the hunter and wrapping in the gene that codes for testes that produce a high volume of T, which in turn drives all the traits needed by man the hunter, Epstein's story about athleticism is a closed loop that revolves around men. We'll have more to say about sex, T, and athleticism in Chapter 7, but we point to this story to underscore the ubiquity of the "man the hunter" story and T's presumed role in supporting it.[13]

Investment Challenges

The challenge hypothesis explains fluctuating T as a matter of changing priorities over time: men can invest time and energy either in getting mates and producing more offspring (mating effort) or in tending the relationships they already have with partners and children (parenting effort). With the challenge hypothesis, T's story is no longer a simple tale of boy meets molecule, and all things manly follow. Instead of looking at higher T as the adaptation that enables male fitness and lower T in dads as a matter of concern, studies based on the challenge hypothesis suggest that flexibility in T increases fitness by helping men to successfully navigate different life stages.

The trade-off between reproduction and parenting is often alternatively framed as a trade-off between quantity and quality of offspring. In a classic 1972 paper, the biologist Robert Trivers proposed that parental investment should be defined as "any investment by the parent in an individual offspring that increases the offspring's chance of surviving (and hence reproductive success) at the cost of the parent's ability to invest in other offspring." For evolutionary theorists, reproductive fitness inheres in the survival of someone's entire genetic lineage, not just that person's direct offspring. Investment is not merely a metaphor in this literature but reveals capitalist economic theory as the foundation for many theorists' understanding of how resource problems are solved in the natural world. For instance, in a theoretical essay that reads in parts like an economics primer, the anthropologists Kim Hill and Hillard Kaplan describe how reproductive fitness is ultimately a function of "embodied capital," which in turn "can be divided into stocks affecting the ability to acquire the resources for reproduction and stocks affecting the probability of survival."[14]

Traditional hypotheses about T and behavior mostly relate to what would fall under "mating effort" in this theory, meaning anything that increases men's opportunities for sex with fertile women. But for humans and other animals whose young have long periods of dependency, ultimate fitness requires something different from ongoing reproduction. And here's where constant high T as an "adaptive male trait" gets tricky, because high T is commonly held to be great for things that help men get sexual partners, but problematic once there's a baby in the picture. "Testosterone promotes egocentric choices and reduces empathy," goes a typical proclamation. Reduce T and Dad is less focused on his own interests and thereby more able to tune in to others' needs.[15]

Put another way, men need a flexible physiology to support roles that vary across the life course. If reproductive success for men doesn't end after insemination, but requires that they provide some direct care for children, then their T must drop. Otherwise, the thinking goes, they will be too dominant, aggressive, and focused on sex to be good caregivers. As Gettler and his collaborators have explained, "Given that testosterone motivates behaviors related to finding and competing for mates, perhaps this social drive was no longer needed, and might even be a distraction, as a male's duties shifted to caring for dependent young." Once they're in that relationship with a partner and especially with a dependent infant, their caregiving interactions stimulate the necessary drop in T. The more engaged a man is, the more his T drops. Traditional thinking holds that high T is the adaptive trait for men, but the challenge hypothesis suggests that the evolved trait is instead flexibility and responsiveness in T levels.[16]

But there are a few problems in extending the challenge hypothesis to humans. First, the hypothesis is fundamentally about seasonal changes in behavior that are linked to parallel changes in T in species that have distinct breeding seasons. Humans obviously don't have a "breeding season." Second, while there are pretty consistent data in humans suggesting that competition, especially physical competition, causes T to rise, there is not much evidence suggesting that higher T makes people more competitive. Nor does higher T contribute much, if at all, to other human behaviors and traits that are considered components of "mating effort": aggression, dominance, and most importantly, sexuality.

Perhaps the most persistent fallacy covered by the idea that high T increases men's "mating effort" is the notion that there's a strong relationship between T and human sexuality. This is probably one of the reasons that James Dabbs and other important T researchers have focused so much on aggression. By the early 1990s, so much data had accumulated to confirm T's minor role in sexual functioning in healthy men that Dabbs could confidently write, "Testosterone appears to be related more to aggression than to sexual activity." If aggression is the benchmark, then T's effects on sexuality must be low indeed; even the researchers who persistently hold on to the idea that T "potentiates" human aggression acknowledge that the effect of T is weak and inconsistent and that the data remain "inconclusive."[17]

Though T is considered important for basic sexual function, only low levels are necessary. A fairly recent study in the *Journal of Sexual Medicine* suggests that it's not clear whether androgens are necessary for penile erection. Sexual desire has been even less consistently correlated with T. As van Anders put it in 2013, "In healthy men, research is quite clear, and completely in contradiction to most general assumptions: T is not significantly correlated with sexual desire." For both men and women, there seems to be a threshold effect. If T is below the lower range of "sex-typical" values, sexuality suffers, but above that relatively low threshold, more T doesn't seem to make much difference.[18]

While challenge hypothesis researchers often suggest that T drops when people are in stable partnerships or become parents as a way to tamp down their interest in sex, population-level data on human sexuality triangulate with individual-level studies of T to undermine that idea. Global data from multiple studies covering fifty-nine countries converge to indicate that, as Peter Gray and Kermyt Anderson put it, "a robust feature of human sociosexuality is that married men (and women) engage in sex more than their single counterparts do." Gray happens to be one of the first people to establish that men in committed relationships have lower T than their single counterparts.[19]

So much for high T increasing the mating side of the life history trade-off. But selective citations in this literature bolster the claim that lower T will deliver dads benefits that include reduced aggression, less urge for dominance, and less randiness without ever noting that these "benefits" not only are speculative but run counter to nearly all empirical evidence about what T does, and does not do, in men.

Racing and Classing Parental Investments

Theories of parental investment and life history strategy are nestled within a broader scientific literature comparing reproduction and sexuality across human populations. The language of "investments" and trade-offs between quality and quantity of offspring should already give a clue about where this goes, suggesting that in some groups parents care for their children and in other groups people just indiscriminately reproduce. There is a strong resonance between that idea and racialized and classed narratives of parenting and sexuality. In fact, this is the exact underpinning of the nineteenth- and twentieth-century eugenicists' fears of "lower" races and classes out-reproducing Europeans. Similarly, against a backdrop that pathologized mother-headed households, twentieth-century scholars used evolutionary theory to scrutinize the emerging "reproductive strategies" of children reared in "father-absent" families. By the time the challenge hypothesis came on the scene, there was already a long-standing research narrative about parenting that had strong racial and class dimensions.[20]

To be clear, most studies of parenting and T are done by researchers who do not deliberately racialize parenting in their studies; anthropologists, in particular, generally don't even frame their subjects in terms of race. Instead, they explore T dynamics across or within populations and ethnic groups, where the term "ethnic" is used to refer to specific cultural and linguistic groups that might have distinct patterns of pair bonding and parenting. They use the term "population," on the other hand, to refer to extremely large, generally geographically based groups, such as "non-Western," "Western," or "North American" men, or to groups differentiated by the prevalence or legal status of polygamy. While none of the anthropological studies we found explicitly suggest that ethnic, regional, or other population patterns for T map onto "racial" groups, general trends in scientific concepts on race and discursive slippage between "population" and "race" might encourage such readings, especially among those readers who are inclined to think of "ethnic" as a more modern or acceptable word for race. Ramya Rajagopalan and colleagues have pointed out that the concept of a population, understood as a "bounded, genetically differentiated group," "permit[s] ideas of race as biology to persist and sanitize[s] the study of human genetic differences in an attempt to diffuse some of the post–World War II anxieties around race."[21]

To examine the subtle slippage between study specific groups and larger populations, consider Martin Muller and colleagues' 2009 comparison of testosterone and fatherhood status in two geographically proximate peoples in Tanzania, the Datoga and the Hadza, the latter of which is widely reported to be the last group of nomadic foragers in the country. After a close comparison of the two groups, the researchers zoom out to consider how their findings might fit into broader global patterns that show T is not reliably related to partnership and parenting in "non-Western populations": "Cross-cultural variation in testosterone responses to marriage and fatherhood could potentially result from differential patterns of investment in mating and parenting effort, even among married fathers. . . . If North American men typically invest more in marital bonds and paternal care than men in, for example, polygynous societies, then this might account for the more predictable association between reduced testosterone and fatherhood in these populations." Here, North America is implicitly figured as a unit that can be meaningfully described as a "monogamous" population and can be contrasted to polygynous ones. But a review of global data on sexual behavior indicates that North American and European adults report more sexual partners than adults in the rest of the world, while regions where polygamy is more common and is legally allowed, especially parts of Africa, show that both women and men have significantly fewer partners than in officially "monogamous" regions.[22]

The ostensibly neutral fact that fatherhood is more often associated with lower T in Western than in non-Western samples gains more of a valence when it is paired with the explanation of what a diminishment in T might do for men and their families. Like other challenge hypothesis researchers, Muller and colleagues use selective citations on T and human aggression to build their case that a reduction in T is an adaptation. For instance, they skip over the fact that research indicates at most a very weak and inconsistent connection between aggression and T, and instead emphasize some recent lab-based studies that suggest administering exogenous T slightly increases sensitivity to threats or social challenges. Muller and colleagues say that increased sensitivity to threat might increase "reactive" aggression, which "could prove costly in the context of childcare, not only because it might involve men in aggressive interactions with other men, but also because it could potentially lead to child abuse, such as infant battering." They acknowledge there are no data linking any form of T to child abuse, but mention a single study that linked higher T with spousal abuse. Like other research that challenge hypothesis researchers use to indicate the sorts of problems that higher T levels might pose for fathers, that abuse study considers T only as an independent variable, affecting behavior instead of being affected by it. There is a subtle opportunism in calling on T as responsive only in the context of parenting, while narrating the purpose of that responsiveness by reverting to the old-school model where T drives behavior. The effect is to make high T levels in fathers seem ominous—though, as we discuss below, many of these same authors are concerned about the normative implications of this research.[23]

Pause for a moment to follow the chain of associations that are built up in this work. "North American" and "Western" men show a predictable relationship between the "parenting" life stage and lower T, while men in non-Western populations don't. Lower T in the parenting stage reflects a life history strategy whereby men are "investing" in higher-quality offspring rather than in a higher quantity of offspring. Lower T shifts "North American" and "Western" fathers' investment away from continued mating effort, while those (implicitly mostly non-Western) men with higher T in the parenting stage may actually be dangerous to their partners and children. These abstractions derived from broad-brush generalization of population differences in T, "mating strategies," and parenting patterns loop back to the same racialized associations found in the aggression/dominance literature: Western, implicitly white men are figured as civilized, loving, nurturing fathers, and generalized "others" are less so.

Researchers themselves might—indeed probably do, given disciplinary tendencies—read the challenge hypothesis itself as neutral, but the language has valence, and that valence resonates with other domains of discourse where judgments are made about "fit" and "unfit" parents and "quality" people. This resonance engages strongly held ideas about race, class, and nation, even though these constructs aren't explicit in most of the studies: they enter as ghost variables. The literature on "father-absent families" that predated the challenge hypothesis helps demonstrate how this sort of resonance works. Challenge-hypothesis-based studies on humans can't be read without regard for the long-established tradition of denigrating the supposed matrifocal black family. Black family structure is pathologized and black fathers are judged for absence (often presumed, rather than actual; often circumstantial, rather than intended) from their children's lives, as though this reflects indifference rather than economic and structural pressures that separate black men from their partners and children. Economically vulnerable men of all races might be uninvolved because of their own lack of resources, rather than lack of dedication to their children. Moreover, studies that assess what men actually do with their children undermine the idea of dysfunctional black families. A 2006 report from the US Centers for Disease Control and Prevention found that, compared to white fathers and Hispanic fathers, black fathers were the most involved with their children on a wide range of activities that included bathing and dressing, feeding or eating meals together, and talking and playing with the child. Black dads came out as the most engaged whether researchers were looking at men who reside with their children or those who don't—the definition, in many studies, of an "absent" father.[24]

At the same time that the narratives and underlying framework in studies of the "biology of fatherhood" are resonant with long-standing ideas about parenting and reproduction in more and less "civilized" people, some researchers in this field deliberately try to intervene in these associations. Perhaps paradoxically, given the valence of terms like "quality" of offspring in the theories they draw on, many researchers in this field even argue that variation in men's physiological responses to parenthood can undermine normative judgments about the best way to be a father. These researchers are trying use the data on T in fathers to elaborate biological approaches to human behavior that aren't essentializing.

Consider the work of the anthropologist Peter Gray, one of the first researchers to demonstrate that pair bonding and parenting were accompanied by lower T in men. Gray has studied fathers in the United States, China, Jamaica, and elsewhere and, with Kermyt Anderson, wrote a book on the biology of fatherhood. Though he studies and discusses the challenge hypothesis, Gray has scrupulously avoided the suggestion that a drop in T signals "better" fathering, or that persistent high T levels equate to dangers for children. In his popular works Gray argues that a father's involvement or lack thereof is neither a simple matter of T levels, nor just a matter of psychological propensities or individual traits. For instance, on the fatherhood blog that he co-writes with Anderson, he recently reviewed a book about "economically vulnerable nonresidential fathers (EVNF)." Gray highlighted ways that these dads are treated in the legal system and have been affected by broad economic forces like the Great Recession, and notes that while these dads are almost by definition "uninvolved" fathers, they generally expressed a strong desire to be more involved with their kids than they are. Rather than seeing them as "deadbeat dads," he suggests, we might view them as "dead-broke dads" whose fathering "limitations are often of ability rather than desire."[25]

But many other scientists aren't so careful to avoid judging dads whose T doesn't drop as less than ideal. One problem is that the language typically used for the challenge hypothesis already has a valence implying that a drop in T signals better parenting: lower T promotes investment in offspring, which in turn enhances fitness, increases the survival of the young, and goes along with prioritizing quality over quantity in offspring. A team of anthropologists at Emory University, for instance, examined fathering styles in relation to testes size, a correlate of testosterone production. They positioned their inquiry within classic challenge hypothesis terms, but also framed investment in children as a choice: "Despite the well-documented benefits afforded the children of invested fathers in modern Western societies, some fathers choose not to invest in their children. Why do some men make this choice? Life History Theory offers an explanation for variation in parental investment by positing a trade-off between mating and parenting effort, which may explain some of the observed variance in human fathers' parenting behavior." And those fathers who "choose not to invest in their children" are likely to be downright dangerous because they don't experience the T drop that "might also both suppress impulsive aggression and promote empathic responding toward a highly vulnerable infant." It's no surprise that the media frequently reports these studies with headlines like "Fatherhood Cuts Testosterone, Study Finds, for Good of the Family," "Better Fathers Have Smaller Testicles," and "Aw Nuts! Nurturing Dads Have Smaller Testicles."[26]

Dads and Cads

While the racialization in challenge hypothesis literature on parenting is almost certainly unintentional, another literature connecting differences in fathers' T to differences in their parental investments centers race in a framework that explicitly biologizes white supremacy. These studies use r/K selection theory, an evolutionary theory developed to explain different species' reproductive strategies, and apply the theory to different human groups. To be clear, the theory itself has no racial content, but it has been misappropriated to argue that human races are essentially "subspecies." Both r/K selection theory and the challenge hypothesis are about balancing available energy and resources to maximize reproductive success, and there are strong parallels between them. In the challenge hypothesis, energy is allocated to either mating effort or parenting effort at different stages in the animal's "life history." In r/K selection theory, the same trade-off is articulated as a species-specific pattern of allocating energy to either high quantity or high quality of offspring.

Proposed in 1967 by the ecologists Robert MacArthur and Edward O. Wilson, r/K selection theory builds on the evolutionary geneticist Theodosius Dobzhansky's insight that unstable or unpredictable ecological niches favor the selection of traits that result in high rates of reproduction, faster development, and relatively low rates of offspring survival. MacArthur and Wilson called organisms with those traits "r-selected." Weeds, mice, insects, and other organisms that have a short lifespan and lots of offspring are common examples. In contrast, relatively stable environments favor "K-selected" organisms, with slower growth and maturation, lower reproductive rates, and higher survival rates. Longer-lived organisms with "expensive" offspring, like elephants, orchids, turtles, and humans, are examples of K-selected organisms. What connects these two models is the notion of underlying investment trade-offs between reproducing a lot of offspring that don't require a lot of energy and producing a few offspring that require more energy.[27]

There's a second thread linking r/K selection theory to the challenge hypothesis, which is that both models connect trade-offs in reproductive strategies to features of the environment: across both theories, in unstable environments effort is allocated to mating, while in stable environments effort is allocated to parenting. In r/K selection theory, the environment that matters is the one that prevailed in the species' early history. In the challenge hypothesis, the environment, which includes social relationships, is presumed to change significantly over the course of each individual's life. For instance, in terms of the avian species around which the hypothesis developed, male birds emphasize mating effort when environments are unstable, but once mating pairs are formed, environments are considered more stable, and the male birds turn their attention to parenting effort.

While r/K selection theory has largely been abandoned because of lack of fit with empirical research, a small group of researchers use it to explore different life history strategies in humans, especially men. In the 1980s, the Canadian psychologists J. Philippe Rushton and Anthony Bogaert wrote a sweeping review that argued, via r/K selection theory, that there are evolutionarily derived differences in patterns of personality, cognition, sexuality, and family formation among human races. Their explicit aim was to advance a biological argument for white supremacy, emphasizing intelligence and sexual behavior. Predictably, they suggest that "whites" are more K-selected than blacks, reflecting a strategy of higher investment in a few quality offspring, while "Orientals" are the most K-selected of the three groups. "Testosterone and other sex hormones," they said, might be a "physiological mechanism" for racial differences in mating strategies. As Celia Roberts observed of a paper in this vein by the psychologist Richard Lynn, the three-part racial ranking by traits and T not only denigrates black men but positions "Asian men as less masculine than white men," demonstrating what critical race theorist Claire Jean Kim calls "racial triangulation," whereby Asians "are racialized relative to and through interaction with Whites and Blacks."[28]

On one hand, papers using r/K selection theory to link T with presumed racial differences in reproduction, sexuality, intelligence, and other traits constitute a marginal (and less than current) literature, produced by relatively few researchers and published in a small number of journals, most notably the *Journal of Personality and Individual Differences* and the *Journal of Research in Personality*. On the other hand, this small literature is connected to relatively influential psychologists who have produced some of the foundational studies on race and T. Within a year of Rushton and Bogaert's essay, the Canadian psychologist Lee Ellis enthusiastically picked up r/K selection theory to argue that "criminal behavior is part of an r-selected approach to reproduction." Ellis examined demographic data on criminality and concluded that people of lower socioeconomic status are more "r-selected." But his greatest emphasis was on race: "Of the three major racial groupings, blacks are the most r-selected, orientals are the least, and whites are intermediate." A year later, Ellis had wrapped "sex hormones" into the package of r/K selection, racial difference, and criminality.[29]

Ellis soon teamed up with another psychologist, Helmuth Nyborg of Denmark, to extend his analysis of racial differences and T. Based on a large dataset of health and social variables among Vietnam-era US Army veterans that had been collected by the US Centers for Disease Control and Prevention, Ellis and Nyborg reported that T levels were higher in black men relative to white men. The resulting paper, which does not mention r/K selection theory, has played an outsized role in linking T, race, and behavior because of the many citations it has garnered. In 1994, Nyborg built on his analysis with Ellis to elaborate how testosterone could serve as the linchpin for Rushton's theory. Shortly thereafter Rushton endorsed this idea, writing, "One simple switch mechanism to account for a person's position on the rK dimension is level of testosterone."[30]

Ellis and Nyborg's paper has reached its biggest audience by far via Allan Mazur and Alan Booth's watershed 1998 article "Testosterone and Dominance in Men." The various stories about T and behavior—dominance or its specific manifestation in aggression; aggression and its relation to crime, sexual behavior, reproduction, and more—all circle back on one another, with evidence about T's relation to one held up as reason to suspect T will be involved in another. Thus, Ellis and Nyborg's analysis of racial differences in T, which was crucial to Mazur and Booth's story about high testosterone and the "honor culture" of young black men, secured a connection between the literature on racial differences in life history/mating/parenting strategies and the literature on dominance/violence/aggression.[31]

Another link in the racialized chain of associations with T repeatedly appears in this literature: racial disparities in health, especially prostate cancer. Sometimes the research on health is couched as evidence for race differences in r/K selection, and sometimes it is just a cover for looking at race and T in the first place. Either way, it's a particularly sneaky move. An ardent disciple of Rushton, Nyborg extended r/K selection theory to link testosterone, supposed racial differences in intelligence, South-to-North immigration, and the "decay" of "Western civilization." Yet while Ellis and Nyborg were both working on r/K selection theory, endorsing and amplifying some of the most noxious and baseless claims about racial inferiority of black people, they piously addressed readers' potential concerns about the "propriety of probing into this sensitive area of research": "We are aware that average racial/ethnic differences in testosterone levels may not only help to explain group variations in disease, but could also be relevant to group differences in behavior patterns, given that testosterone and its metabolites are neurologically very active. While cognizant of the possible misuse of information on race differences in sex steroids, we consider the prospects of beneficial effects to be much greater, particularly in the field of health. Nevertheless, especially in the short-run, scientists should be on guard against even the hint of any misuse of research findings in this area." As historian John Hoberman noticed, "The threat of prostate cancer to black men turns out to have been only a prologue to a more ambitious theory of hormonal effects and the racial character traits to which they supposedly contribute." Ellis and Nyborg's analysis of racial differences in T is a staple in mainstream T research, and it would be worth revisiting their finding, especially given that of all the known covariates of T, especially diet and ratios of lean tissue to fat in body composition, they controlled only for age and body weight.[32]

A process of triangulation among T, traits, and race lends multiple implications to this work, depending on which of the relationships in the triangle is foregrounded: it biologizes race by pointing to a supposed biochemical difference in racial groups; racializes T as the mechanism for differentiating races on "key traits"; and uses racial typologies to support claims about links between T and behaviors.

From Investments to Bonds

While a few researchers employ the challenge hypothesis and other theories about testosterone and parenting to naturalize racial hierarchies, a similarly small but thankfully more influential group of social scientists are using data on T and parenting to explode some of the key categories that are used in T research.

Of all the researchers challenging existing models for T and behavior, Sari van Anders, a psychologist and gender studies professor, probably goes furthest. Like other researchers in this field, van Anders is interested in hormones as proximate mechanisms that facilitate evolved human behaviors, in this case focusing on social bonds rather than life history trade-offs per se. In a major reconfiguration of how T relates to human social behavior, van Anders and colleagues propose that data on T (and other steroids) should be integrated with data on another class of hormones, peptides. Their steroid/peptide theory of social bonds advances a framework that is orthogonal to traditional models of T and behavior, because it does not begin with the familiar behavioral categories. Instead, van Anders and colleagues start with the premise that "neuroendocrine responses [to social contexts] provide a proximate means for addressing evolutionary questions about pair bonds and other social bonds." Backing away from the usual frameworks, they formulate questions that generate room for seeing data in a new light. For instance, they reframe the usual assertion about the trade-off between mating and parenting with the question "Why do pair bonds exist when they limit reproductive opportunity?" Their answer is that "pair bonds may be evolutionarily adaptive, enhancing biparental care and parent-offspring bonds when they promote parent or offspring fitness in some way." The mirroring of familiar challenge hypothesis and life history trade-off theories ends here, as van Anders and colleagues move to a discussion of non-exclusive pair bonds, such as polyamory, and same-sex bonds, signaling their intention to shake up heteronormative assumptions that typically drive this research.[33]

Like the anthropologists who argue that fathers have been involved in direct care of children since early human history, van Anders deliberately undermines the idea that nurturing is "feminine." But van Anders breaks decisively from the "dads" researchers by systematically including women and non-heterosexual subjects and perspectives in her research. It's not that van Anders hypothesizes a similarity of T dynamics in women or those who are same-sex-oriented or polyamorous, compared to the heterosexual men at the center of mainstream theory. Instead, her attention shows that to her, T is not, at root, evolution's proximate mechanism for generating either masculinity or heteronormative coupling. Rather, it's a transcendent, multipurpose hormone that has been adapted for a huge array of uses in virtually all bodies.

To understand the mechanistic connections between hormones and social bonds, van Anders and colleagues endorse the idea that hormones are linked to social contexts via two separate physiological systems, one supporting nurturance and one supporting sexuality. They identify the challenge hypothesis as a foundation for their work, but argue that studies using that hypothesis have failed to consistently link three key behavioral categories with a predicted high or low hormone level: offspring defense, aggression, and intimacy.

Van Anders's conceptualization of parenting is illustrative. Parenting, she notes, includes a very broad set of activities that fundamentally differ according to the needs of the specific situation. "We have this schema—this social construction of parenting—that it's all lovey-dovey and warm," she explains, "but parenting can also involve challenging behaviors, like defending infants, if we think about it cross-species or even when we see our kids getting threatened by someone else." She and her colleagues maintain that the challenge-hypothesis-derived prediction that parenting will be linked to low T is too simplistic and contrary to existing observations, such as that infant defense increases T across several species. They focus their investigations on this and other nodes of evidence where there seem to be paradoxical research findings about what behaviors and contexts are linked to high T, and they resolve those paradoxes by insisting that social cues do not have generic meanings, but that meanings always inhere within specific contexts. Thus perception, gender, social situation, and more become important qualifiers for understanding how specific cues will relate to particular hormones. In one study, van Anders explored the paradox that men's T tends to drop when they become fathers, while other research has shown that men's T rises in response to hearing babies' cries. Her team used a programmable baby doll to create situations where, as in life, the adult's attempts to comfort the baby sometimes stopped the crying, but sometimes did not. Men who heard cries but had no opportunity to nurture the "infant" and those whose attempts at comfort were unsuccessful both had a rise in T, but men who were able to stop the baby doll's crying experienced a drop in T. Ultimately, their reinterpretation of data on context and T leads them to reject the idea that high T mediates a trade-off between mating and parenting, instead arguing that the relevant behaviors supported by high versus low T, respectively, are *competition* and *nurturance*.[34]

Shortly after advancing the steroid/peptide theory of social bonds, van Anders threw down the gauntlet with an article focusing squarely on T's role in social behaviors. "Beyond Masculinity: Testosterone, Gender/Sex, and Human Social Behavior in a Comparative Context" boldly takes aim at the presumption, which she says is "widespread" among researchers, that "masculinity and high T [are] proxies for each other." She goes so far as to label challenge hypothesis studies as emerging from "pre-theory": "Largely based on pre-theory that ties high testosterone (T) to masculinity, and low T to femininity, high T is mainly studied in relation to aggression, mating, sexuality, and challenge, and low T with parenting. Evidence, however, fails to support this, and the social variability in T is better accounted for by a competition–nurturance trade-off as per the Steroid/Peptide Theory of Social Bonds."[35]

"Beyond Masculinity" emerges from van Anders's signature work at the nexus of gender studies and empirical social neuroendocrinology. She takes other researchers to task for a number of methodological and conceptual habits, especially the gender composition of study samples: studies of parenting overwhelmingly focus on women, but when T is involved, they switch to men. Likewise, dedication to the idea of T as masculine has driven the classification of behaviors seen as relevant to T, though, as noted, the hypothesized links (such as "high T connected to aggression" or "low T connected to parenting") are poorly reflected in actual data. She takes special aim at the way aggression has been studied: "T and aggression are rarely studied in women, likely because pre-theory about masculinity precludes this possibility. Pre-theory may drive the continuing quest for correlations between aggression and T in men despite broad null findings." In the end, "this gendering of research subjects and topics" makes it difficult to investigate some key questions, and interpret existing findings. For instance, T and the peptide vasopressin are typically studied in men only, while oxytocin is more frequently studied in women, which makes it hard to compare the functional significance of these chemicals. She is similarly critical of the basic interpretive frame for research that posits that the drop in T often observed among fathers reflects an adaptation in men: how can researchers know this without also studying women?[36]

In the end, van Anders's work demonstrates a radical way forward for researchers hoping to integrate theories on sex/gender across the humanities/social science/natural science divides. She tackles some of the stickiest problems head-on, including how to do comparative work across species while also holding firm to the notion that social constructs shape the context of all human behaviors. She offers a model for employing evolutionary theory in a way that does not assume a heteronormative bifurcation of complementary male and female psyches and neuroendocrine patterns. Finally, she shows that it is possible to explore systematically the link between T and human behaviors without naturalizing gender differences.

While van Anders's work is novel because of her explicit move to break the link between T and masculinity, other researchers are also making moves to interrogate commonsense behavioral constructs like "parenting." In the case of Lee Gettler, van Anders would no doubt note that the inquiry is limited by its exclusive consideration of men. Nonetheless, Gettler's DADS (dedication, attitudes, duration, and salience) model integrates multiple explanatory scales—evolutionary, developmental, ecological, and cultural—and focuses on fathering not as a set of objectively observable activities but as a meaning-laden process. In this context, Gettler asks what the drop of T in fathers really means: "At an intracultural level, we do not know the mechanism by which becoming an invested, caring father downregulates testosterone in some cultural settings (or whether the mechanism varies across cultures). Is it sensory, such as exposure to cues from one's pregnant partner or infant? Is it cognitive, reflecting the mental processes of fathers developing paternal identities, forming social-emotional bonds with their children, and/or accommodating other psychosocial demands of parenting a young child? Does it reflect a shift in fathers' status or social interactions within the broader community?"[37]

Gettler's approach converges with van Anders's in that both are unusually attentive to the importance of social context in specifying the way that particular father-child interactions, such as rough-and-tumble play or holding an infant, will affect the neuroendocrine system. Both also repeatedly note the importance and malleability of gender roles in shaping both parent-child interactions and their meanings, and therefore their neuroendocrine correlates. In trying to elaborate a developmental and cultural model for how the meaning of skin-to-skin, face-to-face human interactions matters, he also makes room for less physically proximate activities to be expressions of a bond. Going beyond the usual dichotomy of father-presence versus father-absence, Gettler integrates macro-level social structures such as political economy and immigration trends on parenting experiences and the meanings that attach to them. For instance, he cites Pingol's argument that "female migration leads to a 'remaking' of masculinities for Filipino fathers who stay behind and care for their children," and speculates that "neoliberal economic policies and heightened female labor participation may have contributed" to the dramatic increase in direct caregiving by Filipino fathers in recent decades. Given that men's employment and family roles were in flux while the young fathers he and his colleagues study were children, he speculates that Filipino men's newly transformed economic and social roles could have affected the development and neurobiology of the young fathers. Though Gettler doesn't frame his DADS model in terms of "bonds," that concept more closely captures the gist of his interest than does the usual frame of "investment" trade-offs. The key is that he, like van Anders, thinks it's impossible to understand how T's dynamism is related to parenting without knowing the context-specific meaning of the interactions that cause T to drop—or not.[38]

Rebuilding the Model T

For people interested in biocultural explorations of human behavior, research that applies the challenge hypothesis to human parenting seems to hold a lot of promise. Much of the research is less reductionist than other streams of T research, delivering on the promise of a biosocial approach that can describe how social experiences and environments become embodied. Multiple researchers in this field are much more attuned to variety in human relationships, culture, and the real histories of life as lived, rather than to the canned and predictable masculine or feminine packages of behaviors and traits that are standard fare in other corners of research on testosterone. It's also worth appreciating that researchers have taken feminist critiques of the "man the hunter" story on board, re-tooling evolutionary stories so that they don't automatically enshrine sex-based division of labor as the linchpin of human evolution. It accommodates new evidence, takes account of women as active contributors to hunting, avoids anachronistic assumptions about tool use, and doesn't treat contemporary foragers as if they can stand in for evolutionary ancestors. Finally, claiming that infant care isn't ordained by nature to be the sole purview of women is music to our feminist ears.

As we've seen, researchers make radically different uses of theories and data on T to explore patterns of parenthood in humans. At one extreme, purported racial variations in T and reproductive strategies are taken up to legitimize white supremacy and argue that human races literally have traveled different evolutionary trajectories. At the other extreme, shifts in T associated with the nurturing components of parenting are used to break the age-old designation of T as "masculine." But it is important to be alert to the background narratives about good and bad parenting that are potentially activated by this work, especially when it is framed in terms of investments in offspring—even when these normative judgments run counter to researchers' explicit commitment to a non-normative approach to human behavioral variations.

It's worth repeating that most of the researchers who use data on T to understand evolution and human parenting in no way endorse the uptake of that material for racist arguments. But resonance doesn't require their active engagement. Synthetic theories about the evolution of human behavior, and about T as a mechanism in those processes, are like the warp and weft of a scientific and cultural fabric. Researchers don't weave the fabric alone—no one could, as the field of data required is too vast. Instead, they must rely on other researchers' work to supply some of the threads that get woven into the overall piece. Those threads, as well as the structure and language of the theory, build racial content into challenge hypothesis work in humans.

Writing about this work presented us with similar challenges. Whether a citation is laudatory or outright condemnation, it underscores the importance of a piece of writing by showing that others have taken it seriously enough to engage with it. Links across studies lend each other mutual support, reinforcing the "fact value" of each through citation. We have opted to write about a number of egregiously racist studies in this chapter, especially, and wish that we could do so without citing them. As scholars, we need better strategies for responsibly identifying deeply problematic work without adding to its fact value.

What do the new data on T and fathers do in the world? In terms of social policy, it says that supporting men to be engaged fathers isn't a new-fangled feminist plot that runs counter to men's nature, but fits perfectly well with what seems to be our human endowment: men have the evolved capacity to be involved parents. Here, then, is evidence that a biological framework for social behavior is not always regressive but can unseat received wisdom about male versus female natures.

But the positive effect of this theory—the message that men are biologically prepared to be involved fathers rather than just sperm donors, that they can and perhaps *should* share early parenting tasks—isn't a freebie. It's still underwritten by the continued message that T must drop for men to take on this role, the corollary being that if T remains high, men will still be caught up in all the nasty stereotypical traits of manliness. They might even hurt the baby.

7

ATHLETICISM

ONE AFTERNOON during the 2012 Summer Olympics in London, we had back-to-back conversations about testosterone, sex, and athletic performance that left our heads spinning. Both conversations were with experts whom we sought out because of our interest in sports regulations that had recently been adopted, restricting the eligibility of women athletes whose natural testosterone levels are considered too high and in the so-called male range. Sports officials claimed that these women have an unfair advantage over women with lower natural levels and so they must lower their levels via surgery or drugs or else forgo competition.

Our first meeting was with a rising young star in the science of behavioral endocrinology who is keenly interested in the complexity and dynamism of testosterone: how it sharply rises in response to interventions like intensive exercise or positive feedback from a coach; how its levels overlap between male and female elite athletes; how people's responses to it vary dramatically. He was amused by the suggestion that testosterone is a good predictor of athletic ability. To him, the new regulations do not make sense: testosterone is important and fascinating, but it's far too simplistic to say that testosterone is the single most important determinant of athleticism.

Several hours later we were at a café talking with the Olympic official responsible for orchestrating their new regulation. In short order and without a hint of doubt, he disputed everything we'd just heard from Dr. Rising Star. In his view, testosterone drops, not rises, in response to training; levels differ sharply between men and women; and, perhaps most crucially, it is the chief ingredient of athletic prowess, pushing men's athletic achievements ahead of women's. The studies that are common sense to the first expert were ludicrous and irrelevant to the second; the second expert's conclusions were "flat-out wrong" according to the first. Both claim that the science involved is not controversial, but their views about how testosterone works in the body are utterly opposed. How could this be? We had fallen into a rabbit hole of scientific debate.

In this chapter, we scan the shape of evidence across a domain in a way that's different from how we've worked in other chapters. Instead of diving deeply into a few studies, examining how their specific methods set up the conditions for producing particular associations with T, we hold studies up against one another to see how their findings resist integration. To do so, we largely take the evidence from any single study at face value. That approach comes with a risk: it might make this set of studies look more solid and less contingent, study by study, than the work in other domains that we've considered. That's not the intention. These studies are characterized by many of the same limitations as work on risk-taking, aggression, power, or anything else. Some of these studies are certainly better and methodologically stronger than others, but we are purposely being somewhat naive about the merits of individual studies here to make a different kind of point: different facts about T emerge from different contexts.

The Miracle Molecule of Athleticism?

If you follow sports, the prevailing idiom alone (think "testosterone-drenched locker rooms") could make you think that T is the miracle molecule of athleticism. A skeptical BBC radio host captured the general mystique of T well when she couldn't seem to believe our claim that you can't predict athletic performance by knowing someone's T level, exclaiming, "I thought we knew that raised levels of testosterone in any human being lead to a better athletic performance!"

This apparently simple statement mixes together several different ideas, including the notion that athleticism is a kind of master trait that describes similar characteristics in different athletes, that "athletic performance" across different sports generally requires the same core skills or capacities, and that T has a potent effect on all of them. Nothing about T's relationship to sports performance works in these ways, but even people who should know better sometimes fall back on this popular story. Douglas Granger, a psychologist and behavioral endocrine researcher who was interviewed for a story on T and sports in the *New York Times,* swatted away the most obvious versions of T mythology: "Steroids are not going to take someone without athletic ability and turn them into a star athlete, or teach you how to swing a bat and connect with the ball." But he continued with a vague, almost magical appeal to the idea that T can "elevate" someone who is already athletic: "If you have a certain athletic presence, testosterone could take you to the next level."[1]

It's easy to see why people get confused. Controlled studies in men show that supplementing natural T with exogenous T builds skeletal muscle mass, as well as some aspects of muscle strength and endurance. It may seem logical to infer, then, that a person with more T will have greater athletic ability than one with less T, but this kind of prediction doesn't pan out.[2]

Studies of T levels among athletes fail to show consistent relationships between T and performance. Some studies do show clear correlation between higher baseline (endogenous) T levels and either speed or "explosive" power, but many other studies show weak or no links between baseline T and performance. Quite a few studies even find a negative correlation, meaning that higher baseline T is associated with worse performance.[3]

Several studies have found that T relates to performance only in specific subgroups of athletes, such as those who play certain positions in soccer or rugby, or players who are stronger to begin with. All of these quick summaries mask a lot of complexity and might also give a false impression, because we tend to fill in the missing information with our expectations. Sports physiologist William Kraemer's team found that T was related to some aspects of performance in Big Ten soccer players, and also found that T was different in players classified as "starters" versus "non-starters." A few details from that study reveal important multiplicities that are glossed over in summaries of the study. First, T was measured from athletes' blood, instead of saliva, but the researchers never specify whether the T measures are for total T, free T, or some other version. Second, T did relate to some aspects of performance, but not the same way in the two groups: in non-starters, T was related to vertical jump height, while in starters, T was related to some aspects of knee strength and flexion. Finally, non-starters had higher T than the starters, which goes against the idea that more T translates to greater athleticism.

It's worth noting that these groups are hardly set in stone—people rotate on and off the bench over the course of a single season, and particularly as they gain seniority on the team. The starter/non-starter distinction gets reified by finding a series of statistically significant differences between the groups, which in turn can make it seem like T makes a more definitive contribution than it does.[4]

Shalender Bhasin, director of the Research Program in Men's Health at Brigham and Women's Hospital in Boston and one of the world's most renowned T researchers, has called out a long history of overreaching on the subject of T and sports. In a 2008 paper, Bhasin's group addressed the apparent paradox that giving people T can increase muscle mass and power as well as maximal voluntary strength but doesn't seem to "build a better athlete." Increases in specific parameters that relate to athleticism don't necessarily translate to improved function. In the researchers' words, "We do not have good experimental evidence to support the presumption that androgen administration improves physical function or athletic performance. Androgens do not increase specific force or whole body endurance measures." In the same *New York Times* article in which Granger proposed that T could "take [an athlete] to the next level," Bhasin sounded a much different note: "The explanations of cause and effect between athletic performance and testosterone are very weak," he declared.[5]

How can Bhasin be so certain that T affects specific parameters like muscle girth or voluntary maximal strength, on one hand, but nonchalantly dismiss the idea that T builds a better athlete, on the other? The key is how you define and measure "athleticism." Just as with aggression, risk-taking, or power, we have to take a closer look at what scientists mean when they study athleticism or athletic performance.

Who Is the "Best" Athlete?

At this writing, Usain Bolt is the fastest human in the world. But he isn't the fastest at every race. In an interview in 2013, when Bolt was asked why he never runs the 800 meters, he responded, "I cannot run the 800, that's out of the question. . . . I've tried it and trained and my PR [personal record] is like 2:07, and that's really slow, like, a woman could beat me." The interviewer laughed him off, saying, "You're going to get in trouble for that!" But Bolt was serious: "It's true, though—they could!" In fact, Bolt was understating the case. Visit Alltime-Athletics.com's list of best women's times for the 800 meters and jump to the very bottom of the list. There you will find a thirteen-way tie for 1,881th place (no, that's not a typo), most recently run by Habitam Alemu of Ethiopia at the 2016 Rio de Janeiro Olympics. Her time was 1:58.99—or roughly eight seconds faster than Usain Bolt's best. In 2018 alone, 498 women ran faster than Bolt's best time, including nearly a hundred teenage girls. It's not just elite women who are faster than Bolt in the 800 meters: the all-time world age best for a twelve-year-old girl is 2:06.90, a record held since 2009 by Raevyn Rogers (USA). It goes against all conventions of sports to say that these women are faster than Usain Bolt. But they are—in the 800 meters. It's just not his race.[6]

Some readers might be tempted to throw the book across the room at this point because comparing 800 meter runners to 100 meter runners looks like an apples-to-oranges comparison. But that's exactly our point: the specific skills and physiologies needed to excel in one sport are not the same as those needed in any other sport, even if the two sports are as similar as running very fast down a track. Our point is to slow down the avalanche of assumptions about athleticism and T, and the related assumption that sex overwhelms other differences between trained athletes. The example shows that sex isn't always the most sensible way to divide athletes, even within a particular sport like running. This might seem obvious when the comparison is sprinters versus marathoners, but it may be surprising that there is such a great difference between specialists in the 100 meter versus 800 meter that even the fastest man in the world can't switch distances and automatically dominate.

Sports geeks love to argue about who is the best, and the fervor to stay on top of the rankings and statistics is almost as competitive as the sports themselves. But even the most detailed metrics can't answer some kinds of questions. Sports encompass an enormous array of activities requiring vastly differing combinations of skills and physical capacities. When is power more important? When is finesse? How crucial is endurance? What about flexibility, hand-eye coordination, communication with teammates, strategy? When you hear "athleticism," do you think of sprinting, where success is all about explosive speed? Or do you think of something like luge, where the ability to isolate body parts in order to make subtle adjustments and to remain flexible and relaxed while going unprotected down a track at ninety miles per hour are just as important as power? And then there are the many synchronized events, like swimming and diving, where individual execution is no more important than precisely matching the movements of your teammates. The idea that there is one core ingredient in the magic sauce for every conceivable sport is frankly absurd.

There are still more twists and turns to consider. Consider strength, which at first blush probably seems like a singular facet of athleticism. But strength is not generic. Holding a handstand takes one kind of strength; sprint cycling takes another; powerlifting takes yet another. Sports scientists and trainers mostly talk about four classic kinds of strength: "endurance," "speed," "maximum," and "explosive." The American Council on Exercise, one of the organizations that certifies personal trainers, adds "agile," "relative," and "starting" for a total of seven types of strength. Agile strength, for example, is "the ability to decelerate, control and generate muscle force" in multiple planes. Different sports require different strengths and hence training regimens. What does the person want to *do* with their strength? What strength does their sport require? Competitive weightlifters want "maximum strength," which is the absolute heaviest amount they can lift once. To develop this strength they need to lift heavy weights for a low number of repetitions, alternating with rest time between lifts. Shot-putters, on the other hand, need explosive strength. They need to propel light-to-moderate weights as far as possible. Sprinters need speed strength. They work toward producing maximal force during high-speed movement, but there's controversy about the best way to train: most sprinters use weight training, but some great sprinters barely lift weights. Strength is not just a contributor to athletic performance but also a result, and obviously there are a lot of components that go into that result.

The problem with trying to flatten athleticism into a single dimension is illustrated especially well by a study comparing three groups of men: elite amateur weightlifters, elite amateur cyclists, and physically fit men who didn't regularly participate in sports. The researchers looked at several aspects of strength, power (which includes the element of speed), endurance, and hormones (both T and cortisol), and found interesting relationships between T and athletic capacity, as well as T and the type of sport or training regimen the men followed. But before we get to these results, let's stay with the question of which group, among the three they compared, are the best athletes.[7]

The weightlifters had bigger muscles and were much stronger and more powerful than the others: in leg extension tests, they reached much higher maximum weights and could move their own maximum loads faster than the other men. If maximum strength is the measure of athleticism, then the weightlifters win, hands down. The cyclists weren't even stronger or more powerful than the non-athletes. But there was another test of athletic performance: cycling workload. This test measures how long a cyclist can maintain a pedaling speed against a given resistance. The researchers also incrementally increased cycling resistance to observe how this affected the men's blood chemistry, heart rate, and perceptions of exhaustion. This time the cyclists turned the tables on the weightlifters, achieving 44 percent greater workload than the weightlifters. What's more, when body mass or the size of the thigh muscle was taken into account, the cyclists were again by far the best, and even the control group did significantly better than the weightlifters. In other words, different measures of strength put different groups on top. In perhaps the biggest surprise, those with *lower* muscle power achieved the *highest* maximal workloads.[8]

This might seem like a ridiculous way to look at this study. Of course weightlifters can lift more weight and cyclists are better at cycling. (Though the non-athletes out-cycling the weightlifters might have been a surprise.) But let's move from the question of who is the best athlete to consider what the study says about T. What if we simply told you that the endurance-trained athletes, the cyclists, had significantly lower T levels than either weightlifters or non-athletes? And then we told you that the cyclists were less powerful than the weightlifters? That would be true. But framing it in this simple way might lead you to conclude that higher T leads to higher strength across the board, and that conclusion would be wrong. Among the athletes, T was positively related to some kinds of strength (maximum and power), and was negatively related to another kind of strength (endurance).

T's Contingencies

When we first started researching and writing about T and sports, we felt sure that there would be a solid bedrock of facts about the things that T definitely does for athletes. But the deeper we went into the data, the fuzzier everything looked. T facts, like all facts, are contingent and are true only in specific contexts. Contexts for facts can be narrow or wide, and it seems that the contexts for facts on T and athleticism are very narrow.

Of all the parameters of physiology that are relevant to athletic performance, the two for which there are the most abundant and convincing links to T are skeletal muscle mass (also sometimes called lean body mass) and physical strength. And then there is the head game: the link between T and competition. More scattered data link T to endurance, hemoglobin, VO_2max, and other variables. We focus on the three strongest areas to give a taste of why general statements like "T builds muscle" or "T makes you stronger" are always partial, problematic, and—in some circumstances— outright wrong. In what follows, we show how some common conclusions about T and athleticism are created by erasing the detail and texture in the research.

T Builds Muscle, But . . .

Testosterone is anabolic, meaning it is a catalyst for building more complex tissues such as muscle out of simpler building blocks such as protein. Some in the scientific community apparently resisted this now accepted fact about T for much of the twentieth century, because while animal research indicated that T had anabolic effects, early studies in humans were inconclusive. As late as 1984, an official position statement from the Endocrine Society "declared that when diet and exercise levels are controlled, androgens do not increase muscle mass or strength."[9] At the same time that scientists were convinced that T wasn't anabolic, bodybuilders and competitive athletes who were taking T were just as convinced that it was.

Shalender Bhasin resolved that controversy by running a study that essentially amounted to a controlled trial of doping. Bhasin randomly assigned forty-three healthy young men to one of four groups. He gave extremely high doses of T (600 mg weekly) to half the men, and placebo to the other half. Each group was further divided into one that exercised and another that didn't. Though the study was small, the finding was clear and strong: compared to the men who got placebo, the men who got a high dose of T developed bigger muscles and got stronger, and the effect was especially pronounced among men who not only received T but exercised.[10]

That classic study is the go-to citation for evidence that T builds muscle. But it's also a great study to look at to understand some of the limitations of that claim. First, to find the effect of T on muscle, Bhasin and colleagues had to give huge doses of T, six times the amount typically given to men for hypogonadism and three times more than had been studied in previous research on the effect of T on muscle. Second, even at these high levels of T, the significant increase in muscle size, and especially in strength, was mostly confined to the group that exercised regularly in addition to receiving T. T alone didn't do much.

This study can't easily be reconciled with another, also solid, study on the relationship between T and lean body mass. Lee Gettler, whose work on fatherhood we discussed in Chapter 6, has also studied T, physical activity, and physiology among men. In 2010, Gettler and colleagues weighed in on emerging suspicions about a bias in most T research— namely, the problem that it has mostly been done in North America and Western Europe. In Western men, studies consistently find that higher T is related to more lean mass relative to fat mass. But among a large cohort of Filipino men, Gettler and colleagues found the reverse: higher T was related to a lower ratio of lean mass to fat mass. While those results contrast with research in North American and Western European populations, they align with other research on men in subsistence-level populations: that is, when available calories are low, men with more fat, not less fat, have higher T. Gettler and colleagues interpret their results in terms of a context-specific model of how the body devotes its precious energy resources to different sorts of tissues, depending on the person's life circumstances. Lean mass is metabolically costly—meaning that it takes more calories to maintain muscle than to maintain fat. Everybody needs some muscle, but how much? Based on their study and research in other subsistence-level or near-subsistence-level populations, these scientists reasoned that, "given the metabolic costs of muscle mass, it would be maladaptive for males to indiscriminately maintain lean mass beyond the level of physical demand."[11]

Maybe the difference is because of cross-cultural variation, or maybe it's because Bhasin's study used high doses of exogenous T, while Gettler's study looked at correlations with men's endogenous T. Either way, the statement that "higher T makes men more muscular" needs at least one caveat to be accurate: either "high *doses of exogenous* T make men more muscular" or "higher T makes men more muscular *in North American and Western European populations*." As a great example of T's multiplicity, there is no generic T that is doing the same thing to muscles everywhere.

T Makes You Strong

Naomi Kutin is strong. She lifts raw, a category of powerlifting that allows none of the equipment like shirts, wrist wraps, or lifting suits that allow a competitor to lift much more weight. Kutin broke onto the scene at the 2012 Raw Unity weightlifting championships, having previously only lifted at home. Standing just four feet eight inches tall and weighing under 97 pounds, the ten-year-old Kutin set a new world record for raw squat in her weight class by lifting 215 pounds, smashing the previous record—held by a forty-four-year-old woman from Europe—by six pounds. The following year, looking like a skinny preadolescent at just 95 pounds, she squatted 226 pounds, again breaking the all-time squat record in her weight class. She then bench-pressed 94.8 pounds, breaking the 97-pound open-class world record, and followed that by deadlifting 231 pounds. It came to a total of 551 pounds. To put this in perspective, let's say you weigh 175 pounds. To squat roughly 2.4 times your bodyweight, as Kutin does, you'd need to lift 420 pounds. This is a phenomenally strong prepubertal kid.[12]

When she started breaking records, Kutin was petite, with no visible muscles. This isn't unusual, once you think about it. Bodybuilders, for all their glorious muscle, aren't known for phenomenal strength, and powerlifters often don't have much muscle definition. Now a teenager and in a more competitive weight class, Kutin is still setting records. She's taller, but she still doesn't look the part of a powerlifter. In a documentary, she says with a little laugh, "A lot of people, when they first find out I'm a powerlifter, they really just don't believe me, because, I mean, I don't really have such visible muscles, and they're like, 'No, I don't think that's right.'"[13]

If you came to Naomi Kutin's story without hormone folklore, with only a dispassionate review of the data on T and athleticism to guide you, there would be no reason to suspect that her success is due to T. Prepubescent girls, in particular, tend to have extremely low T levels. And recently, because of her wins at major competitions, she's one of the few teen weightlifters who have been drug-tested. She's clean. But when we have discussed Kutin's strength with a number of smart people who are knowledgeable about sports, including the sports editor at a world-class newspaper, we have been surprised to find that T is constantly raised as a probable explanation. It's the converse of "innocent until proven guilty": T is credited unless you can prove its absence. To us, this illustrates the tenacity of the idea that T is the master molecule of athleticism. The chance of it explaining Naomi Kutin's amazing strength is almost nil. And there's a much more plausible explanation at hand. Kutin has a powerful nature-nurture combination: she began lifting at seven years old, under the tutelage of her dad, Ed Kutin, who is himself a record-breaking powerlifter.[14]

A study of teenaged Olympic weightlifters suggests that the best predictor of strength might be lean body mass, which has a complicated relationship to T. Among girls in the study, the first analysis showed that body mass was the only significant predictor of weightlifting performance, and T was a predictor of body mass. But, counterintuitively, once the investigators controlled for the girls' size, they unmasked a strong *negative* relationship between T levels and performance. In other words, girls with *lower* T lifted more weight. The research team called their evidence consistent with studies of adult women weightlifters. Those studies, which haven't controlled for body mass, show performance is not related to T level. Controlling for body mass, the researchers said, would likely reveal a negative relationship between T and performance in adult women weightlifters, too. Among boys in the study, the first level of analysis didn't show any relationship between T and performance, but DHEA, the testosterone precursor, was linked to better performance. Again, though, as with the girls, controlling for body mass changed the picture. Among the boys, once body mass was taken into account, there were no significant relationships between any hormones and performance.[15]

The researchers struggle to explain these findings. In the end, they suggest that T (and other steroids) affects multiple body systems, and the relationships sometimes work in a positive synergy toward the kind of strength needed for a particular task but sometimes might work in opposing directions. For example, T affects tissues that include muscle, visceral fat, and breast tissue, among others. While muscle is crucial to force, T also affects fat localization in the lower limbs, which they point to as especially important for certain powerlifting moves. They relate this observation to differences in the specific relationships they found among hormones, body composition, and strength in boys versus girls. Their general point is the importance of context and specificity in understanding how T and other steroids relate to performance, including where in the body T has particular effects, which bodies are affected in particular ways, and which part of the body is most important for the specific sport.

While the study on young Olympic weightlifters is surprising from the popular perspective that T makes you strong, it's actually right in line with the sports research on T. Many studies show that endogenous T is related to strength, but sometimes that relationship is found only in subgroups, like older men, and frequently gets much smaller or even goes away entirely when other known correlates of strength like age, body mass or dimensions, and training are controlled. Together, the research provides strong contrast to the generic idea that T is a simple and overriding ingredient for strength.[16]

In 2013, we sat down with a rower who brought home gold medals from two different Olympics. Rowing is a power sport, and she is neither big nor naturally strong. That left her only one hope for making the team: "You can't do anything about your height, you can't do anything about your size, you just need to get stronger because with rowing, it's a strength-weight ratio pretty much." So she focused on building her strength, taking up CrossFit and consistently going above what her coach directed. She got to be one of the strongest rowers out there by training so hard that several times she endured stress fractures just from the force she was putting on her own body as she pulled through the water. She figured, "If I can suffer more than the next guy, then I can win, and that doesn't have anything to do with your physiology; that just has to do with your mental state of mind." But her strength *is* about her physiology, it's just not about T—her T levels are actually below normal. It is possible that intense training has caused her T to drop, but that doesn't negate the fact that she's still very strong even with her low T. The simple lesson we can draw from her story is that there is more than one route to strength.

Biology is full of examples showing that the same outcomes can be arrived at through different pathways. Take the apparent paradox that men and women get the same relative benefit from resistance exercise, in spite of the fact that T improves the effect of resistance exercise and women generally have much less T. A Brazilian team of researchers recently confirmed a more specific case of the general pattern: with resistance exercise, women build upper body strength as quickly as men do. This might be considered a surprising result, given that T is involved in muscular response to exercise, men's T levels are generally considerably higher than women's, and animal and human studies suggest that there is a higher concentration of androgen receptors in the upper body musculature compared to the lower body musculature. The researchers suggest that the hormonal mechanisms underlying women's and men's strength gains might not be the same: the anabolic effects of progesterone might be making up for women's relatively lower levels of T.[17]

Hormones don't work in isolation. When researchers don't find expected relationships between T and strength, they point to the known contingencies in how the body uses T, as well as other steroids in the related chain of compounds of which T is just one part. For instance, reflecting on why their study of young Olympic weightlifters didn't find even the small or spurious relationships between strength and T (or some related parameter like the relation between T and sex hormone binding globulin, SHBG) that prior reports have found, Blair Crewther and colleagues had two main suggestions. First, the hormone to pay attention to might be not T but DHEA-s, which is the precursor not just to T but to many other steroids. Second, steroid effects "might depend on other endocrine features (e.g., receptor interactions, cell type, hormone degradation, binding proteins)."[18]

Recall that the endocrine system is characterized by positive and negative feedback loops and complex chains of chemical conversions. For example, when T is relatively low in adult men, T acts on the hypothalamus and pituitary to release hormones that act on the testes, which in turn produce more T. That's the positive feedback loop. But when T levels get higher, T shuts down the hypothalamus and pituitary, causing T production to stabilize. That's the negative feedback loop. This is one reason introducing pharmaceutical T doesn't automatically raise someone's T level or increase things T affects: sometimes the body works to maintain stasis. That's not to say that it's impossible to override these systems with enough T, just that T's effects on specific systems can be profoundly different—in fact, opposite—when T is high versus low. Moreover, some T might be directly used, but some is also converted to other hormones, such as estrogen. How much T gets converted depends on factors such as how much aromatase, the enzyme that facilitates the conversion, is present. Finally, T's actions are mediated through receptors, which differ in number and sensitivity across individuals, and hormone-receptor interactions themselves are involved in feedback loops that can produce or inhibit more receptors. Knowledge about how androgen receptors mediate T's actions is still scant but rapidly evolving. One fascinating recent development is evidence that there is a big difference in androgen receptor density in skeletal muscles at different locations, such as the head versus neck muscles, shoulders, lower limbs, and so on. The classic expectation is that T will have greater effects where there is greater receptor density, but it's not that simple. Exercise physiologist Fawzi Kadi and colleagues describe location-specific responses (e.g., neck versus shoulder muscles) to factors like exercise and even experience with exogenous steroids: in some muscle groups, these factors upregulate androgen receptors and increase muscle size and strength, and in other muscle groups, exogenous steroids can have no effect or can even downregulate the androgen receptors.[19]

It's not possible to track this sort of multiplicity in studies of T and athletic performance, because existing studies haven't accounted for underlying variation in receptors. But the multiplicity is always there.

Instead of just trying to decipher T's role in strength in a general way, some research teams have begun to think about T as one of a range of bodily resources that we use to build strength. Strength, after all, is a basic functional necessity. It makes sense that we would have evolved redundant systems for ensuring we have adequate muscle strength for all our necessary physical tasks; relying on T to do all the heavy lifting wouldn't be prudent from an evolutionary standpoint.

William Kraemer and Keijo Häkkinen are major researchers in sports physiology and endocrinology from the United States and Finland, respectively. Their studies have shed light on a well-known phenomenon in weight training (aka "resistance training") whereby people with no prior experience often make much more rapid gains than people who have already been training for a long time. Kraemer and Häkkinen's research indicates that the body calls on different strategies for building muscle under different circumstances. When people begin weight training, their bodies don't yet know how to use the muscle they already have. Their nerves don't quickly transmit the information between muscles and brain to initiate and coordinate all the actions necessary to move a load, especially a very heavy one. But with training, the nerves quickly "rewire" to fire in response to demands made on the muscle. These neural adaptations happen fast relative to the slower process of building more muscle. It's only after the neural adaptations have progressed sufficiently that people begin to build more muscle in order to move heavier weights. This is only one of the ways that our bodies use different strategies to solve similar problems depending on our prior experience and the current context. As we max out one strategy (e.g., nerves are already firing to the max), our bodies move on to the next strategy, and the next, recruiting additional resources to keep meeting the demands we place on them. This work shows that hormone regulation for muscle development is just one of many resources that our bodies call on differentially, depending on prior experience. Specifically, hormones become more important to strength when people have already done a lot of strength training.[20]

•   •   •

UP TO NOW, we have mostly been talking about people's endogenous T. When reviewing the mixed evidence about T's contribution to building muscle, we discussed several studies where researchers administered T or placebo. Most of those studies included measures of strength as well as muscle mass, and as we noted, they show that exogenous T increases both muscle size and strength, especially when it is administered in large doses to otherwise healthy men and is accompanied by resistance training. But when any of those conditions are changed—the doses are lower, or the subjects already have low T levels or are elderly or are women—the effects of T on strength are less consistent. The key again is context and specificity. With a team including Shalender Bhasin, whose work confirmed that large doses of T can stimulate muscle development in men, Grace Huang recently conducted a randomized, double-blind trial of T administration in women. As with men, women who received high doses of T (but not lower doses) gained lean body mass and got stronger, but on only two out of five measures of functional strength. Why did high doses of T improve chest press strength, for example, but not leg press strength or grip strength? It probably has to do with the specific demands of particular measures of strength. Elsewhere, Bhasin has explained that "testosterone effects on muscle performance are domain-specific; testosterone improves maximal voluntary strength and power, but it does not affect either muscle fatigability or specific force. . . . Unlike resistance exercise training, testosterone administration does not improve the contractile properties of skeletal muscle." Strength is multifaceted, and T doesn't have a flat relationship across all different kinds of strength.[21]

The findings from studies of T's effects on strength that we've reviewed are statements of averages, and statements about T's effects under particular conditions of testing. Bhasin's statement about the domain-specific effects of T on muscle performance would be more accurate if those conditions were made explicit: testosterone improves *the average Western man's* maximal voluntary strength *as tested in the laboratory,* or testosterone improves *certain measures of* strength and power, *but not all,* in *Western women who have had hysterectomies.* This is not to nitpick, but to point out the importance of signaling the specific conditions in which relationships have been found—especially when there is evidence that the relationship doesn't generalize to other contexts.

So are T and strength related? Yes. How? In complicated ways that researchers don't yet understand, but surely not in a linear and predictable way. The bottom line on strength is that big doesn't equal strong and T affects strength, but not the same way for every kind of strength or every body.

The Head Game

A common belief about the role of T in sports is that high T will make someone a fierce competitor. There are a lot of data on the link between T and competition, but it's exactly the reverse of that common wisdom: competition very often raises T, and seems to raise it only slightly more in those who win than in those who don't. We described this so-called winners' effect in Chapter 5, on risk-taking. Since the 1980s, the conventional wisdom has been that winning works its magic on the neuroendocrine system whether someone wins through skill or is randomly assigned to win by the researcher, but the jury is still out on whether and how much it matters that the competition is physical rather than intellectual, social, or economic. Most of the sports literature endorses the winners' effect as being solidly supported, and most scientists seem to agree that both female and male athletes experience a rise in T in the course of competition.[22]

But this is an area where, once we looked closely at the data, we got some surprises. We started out taking sports scientists and psychologists at their word on this one, because pretty much everyone seems to agree that there is consistent evidence that competitions cause two kinds of rise in T: the rise that comes from anticipating a challenge, and a second boost for those who win. There's a whole subfield of research devoted to studying whether and how just the experience of winning a competition affects T. Quite a few studies suggest that winning a competition raises T—even "indirect" kinds of winning, like that experienced by fans whose teams prevail, or in experimental subjects randomly assigned to win.[23] When we examine the studies ourselves, though, the data don't look so tidy.

We assessed ninety-six studies of competition and T, many of which were included in two recent meta-analyses, and found that these data are usually summarized as more conclusive and consistent than they are. For one thing, individual studies generally begin with a fairly simple hypothesis, but the findings of rising T are often confined to subgroups, or depend upon analysis of additional factors that vary greatly from study to study. There are disagreements about whether the physicality of the competition matters, or whether competing at, say, chess activates a similar hormonal response as does sports competition. So far, the data aren't sufficient to answer that question, because studies of non-sports competition are mostly lab-based studies, and it might be that people are less invested in a competition if it is contrived for research purposes. Physical activity on its own may raise T, so it's hard to attribute the small changes that studies find so far to competition per se.[24]

There is some evidence for a winners' effect, with both women and men who win competitions tending to show a greater increase in T. But the effects of competition on T are highly heterogeneous, meaning that both the direction and the size of any effect of competition on T is inconsistent across studies. Some of the factors that seem related to whether studies find a winners' effect make intuitive sense, like participants' age (since hormone responsivity changes with aging), whether it's a lab-based or real-world study, and the timing of T samples relative to the competition. But one recent meta-analysis showed that the country of the study was related to whether a winners' effect was found. They also found some evidence that competition studies are more likely to be published if they support the hypothesis of a winners' effect.[25]

Recent studies are more likely to look at T in relation to other hormones, especially cortisol. For example, some researchers call on a dual-hormone hypothesis, which suggests that the specific relationship between T and competition depends on someone's cortisol levels. Studies in this vein often deliver findings that T is related to some aspect of competition, but for very specific subgroups. For instance, a team at the University of Chicago recently conducted a study in which 120 women competed against each other on a computer game. The team examined T as both a potential predictor of competitive behavior and an outcome of competition and found that baseline T was associated with the accuracy of performance in the competition, but only among those who lost the competition, and only if their cortisol was low to begin with. They found no evidence of a "winners' effect" and no evidence that competition itself raised T, regardless of cortisol levels. Other researchers appeal to additional complex hormone models, such as the coupling model that suggests T and cortisol may either work synergistically or antagonistically, depending on the context. These studies don't support the idea that higher

T makes someone more effective in the competitions they undertake, even though there does seem to be evidence that winning competitions raises T, at least briefly.[26]

One thing is clear from the studies of T and competition: pre-competition testosterone levels do not predict an athlete's performance on the field. And the winners' effect turns out to be more elusive than most of the researchers currently allow. It's the Mulder effect again: they want to believe. One research group's enthusiasm on this point looped them right back to the popular idea that higher T predicts a win, as they proclaimed that "post-exercise salivary testosterone levels could have the potential to predict performance in endurance running." Unless they are trying to predict *past* performance, that's not going to work out.[27]

Looking Directly at Variability

Up to here, we've shown that for the three domains in which evidence on T and athleticism are strongest—that T builds muscle, that it makes you strong, and that it's linked to competition—links with T are elusive, partial, and contextual. Just like all the other domains we've discussed, from parenting to financial risk-taking, there's better evidence for athletic training and competition affecting T than the other way around. T levels are part of the story, but what someone's tissues do with T is what really matters. People have very different responses to the same amount of testosterone. The studies reviewed so far in this chapter mostly provide group-level data only, which masks inter-individual variability. When investigators find a correlation between T and something like muscle strength in a subgroup (say, those who exercised regularly versus those who only received T from the researchers), they've identified one factor that changes the way T acts in the body. But there are always more factors that they haven't identified, ranging from biological differences between subjects to unmeasured variations in their behaviors or daily environments.

Many researchers who study T in athletes have begun to look directly at the variability in T's relationships with physical and psychological factors related to sport. They examine how specific individuals in their studies are different from others, as well as how individuals' own fluctuating T over time and contexts might relate to their capacities related to sport. These studies produce an entirely different view of T, one that is inherently contingent. It's not easy, and sometimes not even possible, to reconcile the conclusions about T and athleticism that emerge from analyses of how T works at the group level with analyses that focus on individual variability.

•   •   •

BLAIR CREWTHER IS AN energetic and prolific sports scientist whom we met in 2012 at his lab at Imperial College in London, to talk about his T research. Crewther's studies aim at developing interventions to maximize athletes' training potential. He sees T as one of many characteristics that vary among people, and that contribute to but don't determine athletic potential. With T, "some people are simply blessed with lots and others not so much." But he and his colleagues have been experimenting with ways that they could stimulate rises in athletes' T levels, and then see what that does for their performance. On our first meeting, Crewther was bouncing around with enthusiasm about this work trying to boost T through interventions like reviewing a game video while a coach gave positive feedback. "T is extraordinarily dynamic," he explained. "You can see it go up even 100 percent."[28]

When we met with him, Crewther had just finished a small but clever study of rugby players. That study is especially interesting because of the level of detail they provided on each individual who participated. Crewther and his colleagues looked at testosterone, strength, and sprinting speed in ten professional male athletes. Half the men were rated as "average" at leg presses (a bit of an understatement, from a lay perspective, as this involved pressing up to twice their body weight) and the other half rated as "good" (pressing more than twice their body weight). There was no difference in T levels between these groups when they began. After ten training sessions, everyone's T went up similarly, rising 100 percent on average. But only half the men, those in the stronger group, also got faster or increased their leg press weight. The other half—who are also professional athletes, so no slouches—didn't get any additional boost in strength or speed by doubling their T.[29]

There are layers upon layers of variability to consider in absorbing the data on how T responds to exercise. There are fluctuations within individuals, both of T and of other factors like diet, sleep, sexual activity, menstrual cycles, and more that might affect their T production. There are variations between individuals in things like receptor activity, other steroid hormones and enzymes that affect the steroidogenesis chains and feedback loops, and more. And there are variations in research methods, where different teams choose different forms and intensities or duration of exercise, different methods for collecting and analyzing T, and different statistical models for analyzing the relationships.[30]

We've only scratched the surface, and have only focused on those aspects of athleticism—muscle mass, strength, and T's response to competition and exercise—where the evidence for T is strongest. Most simple questions about T don't have clear answers, even in the sports endocrinology literature, including: Does training raise T or lower it? That depends on the type of training, and also on inter-individual variability. Does competition raise T or not? Perhaps, but if so, it may depend on the type of competition, the expectation of the competitor, whether the person wins or loses, and other inter-individual differences. If T levels do rise, are those higher levels associated with better or worse performance? That depends on the type of performance you're measuring, as well as inter-individual characteristics like baseline T, baseline performance variabilities, cortisol levels, and other characteristics that underlie those variabilities. Should T be measured, or is DHEA (precursor to T but also to other steroids) potentially more informative in women? Should samples be taken from serum, urine, or saliva, or possibly from tissues like hair?

The multiplicities go on and on. Total T, free T, bioavailable T, ratios like the "free androgen index," of which T is a part; morning T, evening T, T during the luteal phase of the menstrual cycle; baseline T, T after resistance exercise, T during or after competition—and more. It depends on which T is best suited to the scientists' question and can be fitted into their disciplinary knowledge. For example, clinical researchers almost always measure T in blood, and it would be difficult to suddenly break with that and measure subjects' salivary or urinary T: how would they offer meaningful comparisons to other studies in the literature? But sports scientists often measure T in saliva, not only because it is easier to get athletes to spit into a tube than to draw their blood, but also because some of them believe that salivary T better captures the rapid aspects of testosterone action in the neuroendocrine system than does T in blood.

All the disagreement among studies doesn't suggest that T isn't "really" doing anything, but rather suggest that T is a multipurpose molecule whose specific actions elude our models, which demand a sort of "fungibility" across contexts. T is involved in a lot of body processes and mechanisms, but in highly specific ways. Our attention to these complexities has sometimes been misunderstood as us saying that T does nothing for athletic performance. To the contrary, one of our most important take-home messages about T and athleticism is that the hormone probably does far more than most people realize, but that those effects don't necessarily add up to produce "better" performance, especially not across all sports.

Usable Facts

This deep dive into some of the most well-accepted ideas about T and sports yields surprisingly little in the way of conclusive evidence about what T does, instead showing a potentially frustrating collection of highly specific facts that seem to resist synthesis. This dissonance is what happens when researchers with different basic objectives study T's relationship to physical functions that are relevant to sport. They approach their studies with very different goals, from resolving the question of whether T therapy might counter frailty in the elderly, to learning which strength training programs might naturally boost athletes' T, to seeking evidence that might support an embattled regulation on women athletes. These goals in turn shape the facts that emerge from studies: Who emphasizes group differences, and who reports on individual variability? Do researchers approach T as a trait that might *predict* something about physical competence, or are they interested in how T *responds* to training or competition, perhaps to tailor training programs?

The range and detail of facts that are available about athleticism and T can be overwhelming, partly because without having a specific purpose for the information, there's no way to anchor a search through the data. Free-floating facts aren't the same as evidence, because the latter implies a specific hypothesis or problem. This isn't just about T, but about science in general. Several decades of research across multiple disciplines show that just as science is not unitary, neither are the facts that emerge from specific sciences, or even specific studies. This point is different from the observation that multiple studies addressing the same question yield different results. Some studies show that athletes who compete generally get a rise in T whether they win or not; other studies show that only the winners' T goes up; the bulk of studies seem to show no change. Those studies can be relatively directly compared and synthesized through approaches such as meta-analysis. Likewise, we aren't talking here about good versus bad science. In previous chapters, we've shown some dubious and even outright wrong scientific practices in high-profile research on T.

A good example is Carney and colleagues' work on power posing, where, among other things, they p-hacked, picking their variables after they had already looked at the data. Another is the famous study by Coates and Herbert, where they chose variables that don't actually answer their research question. We are talking instead about the fact that "good" science, science that follows all the rules of good methods, will produce particular facts that don't easily align.

Facts are produced through specific questions, techniques, tools, and interpretive frameworks, and values are embedded in all of these. This is very different from saying that science is "just made up" or "the same as opinion." It means instead that while the material world does indeed exist, we can only know that world through our human engagements with it. The best we can do is use our senses, which allow us to perceive and select only some data points out of all the possible phenomena that exist; transform those data by filtering them through our measures; and apply our own cognitive, linguistic, and disciplinary frameworks to shape the results into an interpretation that is meaningful to us.

Science studies scholars have increasingly appreciated that this entanglement between our research methods and research subjects means that the facts that emerge from different research programs might not just be different or partial, but might actually be irreconcilable. There is a well-worn metaphor about partial knowledge that tells of different blindfolded experts who must describe an elephant, each feeling a different part of the animal. The idea is that each one provides a correct description of the part they can feel, but that each partial description is profoundly misleading to someone who needs to understand an elephant. As we said in Chapter 1, though, we are talking about an even deeper dilemma. The understandings that emerge from different scientific approaches can't just be added up to make one coherent whole. And that's because in the elephant metaphor, the blindfolded expert doesn't transform the elephant: she merely conveys what is there. But science isn't that neutral.

How, then, are people supposed to use evidence to solve real-world problems? How do people who want or need to use knowledge about T for practical purposes decide which science is right for the job? The philosopher of science Helen Longino has coined the term "pragmatic epistemology" to underscore that it's not enough to understand that sciences diverge: we need a solid methodology for choosing the right science for the particular task at hand. Pragmatic epistemology entails clarifying *why* actors wish to know, and what they wish to do with that knowledge. One of Longino's intentions was to explain how equally valid scientific explorations might reach conflicting conclusions. But deciding which scientific approach is right for solving a particular problem is different from pursuing a "scientific" approach that is guaranteed to support a specific conclusion. Instead of pragmatic epistemology, we might call this opportunistic epistemology.[31]

Regulating T among Women Athletes

With regulations limiting the level of endogenous T in women athletes, the International Association of Athletics Federations (IAAF), now World Athletics, and International Olympic Committee (IOC) opened a new chapter in the history of "sex testing" of women athletes. The T regulations, first introduced in 2011, were supposed to be more scientific than earlier attempts to determine who would be eligible to compete in the women's category for elite sports. We and many others have written extensively on the history of sex testing and the science and ethics of rules that single out naturally occurring variations, like T levels, to exclude some women from competition. A lot of what we have written on this topic has focused on the scientific claims, not because we think science is the most important way to evaluate this rule, but because of the way sports officials have framed this regulation. With earlier versions of eligibility regulations, officials frankly stated that their goal was to definitively ensure than anyone competing in the women's category was a woman, believing that science could provide that assurance. But science can't actually answer that question. There are at least six markers of sex—including chromosomes, gonads, hormones, secondary sex characteristics, external genitalia, and internal genitalia—and none of these are binary. Sex-linked biological traits can vary within individuals, resulting in various combinations. Prior attempts that sports governing bodies made to determine sex ran afoul of this complexity.

This time around, officials sought a different way in, claiming that they had identified the biological component of "advantage" that explains the difference in performance between male and female athletes: testosterone in the so-called male range. The new rules say that women whose bodies produce high levels of T must lower it to below a specified threshold in order to compete, unless they can show that they are completely insensitive to T.[32]

Ever since these rules were released, they have been hotly debated in terms of ethics, confidentiality, rights, values, justice, and science. Common understandings of science would suggest that this would be the part that is easiest to resolve: does the evidence support such a regulation or not?

Is There a "Sex Gap" in T?

The idea of a "sex gap" in T is a cornerstone of this eligibility policy. Sports governing bodies conclude that men's often superior athletic performances result more or less directly from their higher T levels. Policymakers have repeatedly stressed that T level is a sharply differentiated trait between men and women, citing the claim that "there is no overlap between testosterone blood levels in healthy men and women."[33]

The only two large-scale studies of T in elite athletes draw contradictory conclusions regarding a sex gap in T. The GH-2000 study, sponsored by the IOC, included 446 men and 234 women competitors in fifteen highly varied Olympic events whose blood was sampled in 2000 at multiple national and international events. The ranges for men's and women's T values among elite athletes in this study were different from those usually seen in the non-athlete population. Most pertinent to the question of the regulation, the study found substantial overlap in the T levels of the women and men athletes, although the mean values were quite different for the two groups. Among women, 13.7 percent had T above the typical female range, and 4.7 percent were within the typical male range. Likewise, 16.5 percent of the elite male athletes had T below the typical male range, and 1.8 percent of them had T low enough to be classified as within the typical female range.[34]

Shortly after that study appeared, IAAF researchers published their own study based on blood that was drawn from 849 women athletes in track and field from the 2011 Daegu World Championships. In that study, just 1.5 percent of women athletes had T above the female reference range, a sharp contrast with the 13.7 percent in the GH-2000 study.[35]

Disagreements about which study to believe have centered on three main issues. First, policymakers and other proponents of the view that T is sex dimorphic have pointed out that the GH-2000 study used immunoassay to analyze the blood, while the Daegu study used mass spectrometry. Immunoassay overestimates T at lower values, while mass spectrometry yields more accurate T readings at lower values. Thus, they argue that the GH-2000 report might have overestimated the T values among women. But critics have responded that this difference in lab techniques cannot explain the fact that a considerable proportion of men had very low T levels; in fact, immunoassay's bias toward showing higher values would obscure, rather than highlight, a trend toward very low T values in men. Thus, the use of immunoassay might mean that the T values reported are overall higher than true values, especially among people who have lower T levels, but it cannot account for the finding of a male-female overlap in the GH-2000 data.[36]

The second disagreement concerns when to draw serum, because of possible T changes in response to competition and/or intense exertion. Policymakers suggested that the female-male overlap in T observed in the GH-2000 data may be an artifact of sampling within two hours after competition. They cite a single study of women and men participants in an Ironman competition to argue that men's T levels drop, while women's levels stay the same or rise modestly after competition. But think back to our discussion about the effect of competition on T levels, which suggests a much more complicated picture. The trends in the data suggest that how T responds to competition seems to be best explained by the type and duration of competition—not the individual's sex. Intense resistance exercise and short-duration exercise are often associated with an increase in T, whereas studies of endurance exercise (especially lasting more than three hours) more often show a decrease in T. The "correct" time to sample T depends on the purpose of the study, but the timing of blood draws seems unlikely to determine whether a study finds that T levels overlap in women and men. But it is worth noting that athletes are typically tested for doping, including with androgens, after they have won high-level competitions, and doping tests are one way women with high natural T are flagged. Thus, understanding how natural T responds to competition is important, and the timing of blood collection in the GH-2000 might be especially apt.

The third point of contention about the GH-2000 versus the Daegu studies is the most fundamental: the rules for subject inclusion and exclusion. Both scientific groups agree that subjects who have doped should be excluded. Where they part is on the question of whether women with naturally high T should be excluded. The two camps take opposite views on whether to include these women—a decision that bears directly on whether their findings support or undermine the idea that T is sex dimorphic. The GH-2000 study included all women with high natural T in the sample. The Daegu study included women with high T of unknown etiology, but excluded as "confounding factors" all women whose high natural T could be traced to variations in sex development. In simple terms, some of their biological characteristics would be classified as female-typical and others as male-typical. Nevertheless, like their competitors, these women were assigned female at birth, and have lived and competed as women their whole lives.[37]

By what logic do the authors of the Daegu report classify women with intersex variations as "confounders" rather than legitimate subjects in the study? In the Daegu report and elsewhere, policymakers consistently identify intersex variations as disorders and health problems requiring intervention. They reason that this justifies excluding women with intersex variations when they generate reference ranges of T among athletes. But women with intersex variations are not necessarily unhealthy. High T can be associated with health issues but is not, in and of itself, a health problem for women. An a priori understanding of women with intersex variations as unhealthy and thus outside normal variation creates a rationale for their exclusion both in reference ranges and in the regulations.[38]

There is both a surface argument here and a subtext where "unhealthy" signals "abnormal," and where T values that can be read as "outliers" suggest that women with such high T are "outsiders." These subtexts can't be settled through scientific argumentation. Still, there is a strong scientific argument for including all women in the sample. These studies aim to establish T reference ranges for elite athletes: that is, the focus is on physiological ranges, not on establishing what T levels are associated with the presence or absence of particular variations or clinical conditions. This approach calls for descriptive statistics, and in this case, there is no valid basis for discarding some values as outliers. In both studies, if the full range of values for women's endogenous T is included, there is an overlap in T. Through this reading, the exclusion of women with intersex variations can be seen as a circular move, because once those women aren't in the dataset, the study finds that women athletes' T levels resemble standard reference ranges, which justifies the interpretation that T levels are dimorphic and justifies the regulations.

Moving outside arguments that are narrowly focused on whom to include and exclude, either in a particular study or in the women's competition category, it is common for sports scientists to acknowledge that women's and men's T levels overlap. Over the past half-dozen years, we've spoken to numerous scientists in a range of fields, including endocrinologists, the longtime head of a national anti-doping lab, a high-profile sports scientist working in professional sports, and scientists who study elite amateurs. All of them matter-of-factly agreed that women's and men's levels overlap: T levels aren't dimorphic, but are characterized by two curves that overlap more or less depending on how you choose the women and men in the sample. The decision about whom to include as research subjects isn't an arcane methodological issue, but a social and ethical one concerning how we understand and frame human diversity.

What Does T Do for Athletes?

There's no question that men, on average, have higher T levels than women do. Policymakers have asserted that men's higher T is the "one factor [that] makes a decisive difference" between men's and women's athletic performances. To support this, they have pointed to a range of research that shows correlations between higher T and stronger performances in a variety of sports.[39]

One of the studies that they have drawn on repeatedly is Marco Cardinale and Michael Stone's 2006 study of the relationship between T levels in blood and the vertical jump height of elite men and women athletes. We mentioned this study previously as an example of evidence that T levels are associated with better athletic performance. And they are—at the group level. The forty-eight men and twenty-two women in their study were elite athletes in a variety of sports. The researchers emphasize a number of group differences, such as "sprinters have the highest T." These comparisons are captured well by condensing all group members, such as athletes in a specific sport, into a single bar (Figure 7.1).[40]

This is one of a handful of studies that report data in ways that make it possible to see both group-level analyses and individual variability. When it comes to showing the relationship between T and jump height, the researchers chose a visual that emphasizes individual variation by mapping athletes' T levels against their jump heights (Figure 7.2). Though a line cuts through the data from the lower left to the upper right to indicate that statistical analysis reveals a linear relationship between T and jump height, the points on the graph are scattered.

Those trained in statistics will look at this line more than at the individual dots, because the line represents the underlying relationship between T and jump height once the "noise" of individual variation is removed. But what if individual variation is precisely the point? What if individual variations aren't "masking" the true effects of T, but T's effects are achieved through multiple pathways, some of which are synergistically

Women

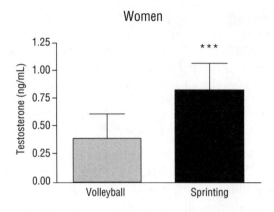

Men

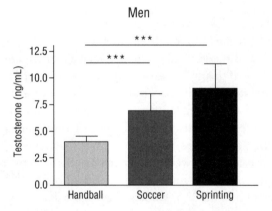

FIGURE 7.1 Comparison of resting serum testosterone level (in ng/mL) by sport for women *(above)* and men *(below)*. Bars show mean ±SD; *** = $p < 0.001$.

(Recreated for the authors by Sheila Goloborotko from Marco Cardinale and Michael H. Stone, "Is Testosterone Influencing Explosive Performance?," *Journal of Strength and Conditioning Research* 20 (2006): 105.)

contributing to jump height, and some of which are irrelevant, and some of which might even detract from it (as by contributing to heavy bones and muscles)?

Instead of focusing on the line, examining the individual data points reveals that some people with quite good jump height have very low T, and conversely, some people with relatively high T are among the poorest jumpers. The man in Figure 7.2 who is at about 8.75 ng/mL on the X axis and about 35 centimeters on the Y axis is in roughly the top 15 percent

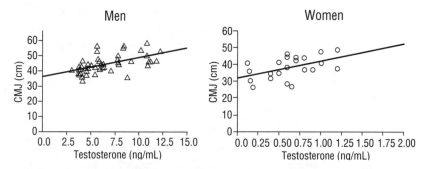

FIGURE 7.2 Scatterplots of relationship between serum testosterone level (in ng/mL) and countermovement jump (CMJ) height for men *(left)* and women *(right)* athletes. For men, $N=48$, $r=0.62$, $p<0.0001$; for women, $N=22$, $r=0.48$, $p<0.01$.

(Recreated for the authors by Sheila Goloborotko from Marco Cardinale and Michael H. Stone, "Is Testosterone Influencing Explosive Performance?," *Journal of Strength and Conditioning Research* 20 (2006): 105.)

in terms of T level and the bottom 15 percent in terms of jump height. Does this person's high T and relatively low jumping ability mean there is not really an association between jump height and testosterone? No. But focusing only on the group-level association—the line that the researchers placed over the data points—can make T seem determinative, instead of being just one ingredient, one that doesn't seem to have the same payoff for everyone. It's not just that you need to have athletic training and experience: everyone in the study is an elite athlete. Likewise, the pared-down conclusions that "sprinters have the highest T" and "sprinters jump higher than volleyball players" obscures the very important variations in how T is related to jump height both across all individuals and among athletes who play particular sports.[41]

How is this relevant for the testosterone regulations? In 2015, the international Court of Arbitration for Sport (CAS) reviewed the science related to this regulation when a young Indian sprinter, Dutee Chand, challenged it. Policymakers leaned hard on the Cardinale and Stone study, saying it "solidified a proven correlation" between T levels and performance, specifically explosive strength. In turn, scientific experts for Chand noted that "correlation does not establish causation," and said that the IAAF had failed to show that T made a definitive difference in performance. The CAS arbitrators split the difference, indicating that they were satisfied with the evidence that T affects performance, but saying that the IAAF did not adequately support its claim that the relative difference between women with high T and their lower-T peers was as great as the average performance difference between elite women and elite men. They suspended the rule and gave the IAAF two years to come up with more data to support it.[42]

•   •   •

STUDIES THAT MEASURE T and performance at a single time point can only show correlations. To answer questions about causation, you need different designs that can isolate changes in T and see how performance is affected. Blair Crewther's work, which we described earlier in this chapter, does that by measuring T and performance before and after interventions that raise athletes' T levels. Recall from his study of professional rugby players that not all players whose T goes up get any performance boost. Some do, but just as many don't. Crewther wasn't particularly surprised by that result, because it accorded with other research that shows little to no direct relationship between T and performance outcomes like strength or speed. In the study of young Olympic weightlifters, in which he and his colleagues didn't find any relationships between hormones and performance, they speculated that other aspects of the endocrine system, including interactions with receptors, binding proteins, and so on, all affect how steroids operate. In other words, steroids don't do the same thing for everyone, and it is outright mistaken to give too much credit to T alone for an athlete's performance. As Crewther says, "You also need to understand the T mechanisms at play (e.g., metabolic, behavioural, morphological, etc.), which will proportionally contribute to athletic performance and adaptability as needed . . . so not all athletes need it, though you probably need 'enough' to function properly, like any natural biochemical."[43]

Policymakers have relied on another small study that includes both individual-level data and changes over time but draws very different conclusions about the relative effects of T. Joanna Harper, a trans woman runner, was noticeably slower soon after beginning testosterone suppression. She requested race times from other trans women runners before and after testosterone suppression, and reported that seven of the eight women were slower after lowering their T levels. She was called as a witness for the IAAF when Dutee Chand's challenge to the IAAF rule was heard at CAS. The court's decision referred approvingly to Harper's conclusion that "testosterone is not the only factor but is the single most important differentiating factor" in athletic performance.[44]

There are many other factors besides T to consider, but for most of these athletes, the only one that Harper considered was age. When the runners were men, they were younger, so the female-male difference is confounded by aging. Examining their age-by-gender rankings for particular races, Harper found that athletes' age-graded rankings remained remarkably similar once they transitioned: they went from "competitive male to equally competitive female." But Harper's examination of the one runner who did not get slower demonstrates a few of the other factors that should have been considered for all runners. Harper explained that "runner seven . . . raced as a 19-year-old male recreational runner and then resumed running years later as a female. She got serious about the sport after she resumed, doubled her training load and dropped 10 kg of weight. Not surprisingly, she got faster." The description is a window into the contingencies that affect athletic performance, but Harper didn't include any such discussion for the athletes whose times followed her hypothesis. Especially given that the training differences and weight loss affected that runner's performance ranking three times as much as her transition did, this is a profound omission. Aspects of this report make it read more like an example of T talk than like a study per se. The data are extremely loose: no T measures; a small, online, convenience sample; and race times that were mostly unverified, achieved on different courses and under different conditions. It's an error to conclude that T is the single most important factor if that's the only factor you look at.[45]

The final point to make about this study is one of opportunistic epistemology. The data were gathered to answer whether it is "fair" for trans women to compete against other women runners. Harper argues that "none of these women has been particularly successful at the highest levels of sport after gender reassignment, and one could argue that this lack of success over ten years would be a strong indication of the fairness of permitting trans women to compete against cisgender women."[46] Unfortunately, Harper took data that she gathered for making an argument about inclusivity and applied them to another case where they would weigh instead on the side of excluding women with naturally high T. Taking her conclusions at face value for a moment, what must one assume to believe they apply equally well to women athletes whose own bodies produce high levels of T? Perhaps most importantly, one must assume that T levels are independently powerful, and that individuals' developmental histories and other elements of their complex neuroendocrine systems are passive in how our bodies use T. But this is wrong. To understand more about how this doesn't make sense, it's useful to turn to another analogy that policy-makers have employed: that of doping with exogenous T.

In debates over limiting women athletes' testosterone levels, regulation advocates frequently use doping as a conversation stopper: doping with T boosts athletes' performance, so athletes with higher T obviously must be advantaged. The idea that T is so overwhelmingly powerful leads advocates of this regulation to the conclusion that high T isn't "fair" even though the situations are fundamentally different: one is cheating and one is not. For us, that's enough to settle the debate and make the regulation illegitimate. But in the context of a book that's exploring how T's interesting and complicated capacities get flattened by facile T talk, it's worth going into some scientific reasons for the analogy's failure.

Martin Ritzén, a Swedish doctor who was on the IAAF medical committee that drafted the regulations and was called as an expert witness in the Chand case, stated in a media interview that doping with T makes an even bigger difference for women athletes' performance than for men, because of different response curves: "If you add a little bit . . . of testosterone to the female low levels you make a lot of difference. If you add a little bit of testosterone to the higher male levels you don't make much of a difference." This might suggest that women are simply more sensitive to T, and that very small amounts of T make a very large difference in women but not men. Applied to the regulation, proponents imply that this justifies regulating women's, but not men's, T: women at the very high end of the distribution will get a greater "advantage" relative to their peers with low T than would be the case with men. But a close look at data from the extensive doping program in the German Democratic Republic suggests something more subtle is going on. A report cited by the IAAF in the testimony it presented at CAS does indicate that doping was especially effective in producing extraordinary performances for women, but it wasn't coming from small doses of T. To the contrary, the T doses given to women athletes "were surprisingly high, and many of the top women in track and field events and in swimming took amounts of androgenic steroids that were higher than the doses taken by male athletes in the same or comparable events." Women's higher doses—sometimes double the amount for men—were mostly used in sprint and hurdle events. They were given ever higher doses, countering the idea that small increases are all that's needed for women's performances to leap to extraordinary levels. Finally, a chapter on doping in a recent endocrinology text suggests that men, too, continue to get increasing benefit from higher levels of exogenous T, and that there is no evidence the effect plateaus. The main conclusion you can draw from all these data together is that evidence for how much T is required to make a big difference in performance is all over the map, as is the question of whether women are especially responsive to T.[47]

There is another reason that data on doping can't easily be used to interpret the effect of natural variation in T levels: variabilities in response to T are substantial. A clinical endocrinologist we interviewed suggested that one reason studies don't always find consistent links between T level and physiological variables is that sometimes high T signals that a person isn't very efficient at using T: the body is producing more precisely to arrive at "typical" function. This fits well with the general biological principle that there are multiple developmental and mechanistic routes to the same outcomes. When it comes to T, a person's full life history is relevant in how any given T level will be available and usable to tissues. Dr. Faryal Mirza suggested that if someone has gone through puberty with very high T levels, the tissues have become "habituated" to those levels, and a lower level may cause poor function. Thinking back to the Harper study, this is another reason to reject it as showing that lower T levels are responsible for the slower times that runners clocked after T suppression. In Mirza's view, it's not the fact that T is low, it's the fact that T dramatically dropped from those individuals' previous hormonal environment.[48]

An endocrinologist who testified as an expert witness for Dutee Chand in her CAS challenge agreed that "there is 'absolutely no doubt' that athletes who have doped with exogenous testosterone have experienced performance benefits." But he said this was a flawed analogy for understanding naturally high T levels, both because "there is no demonstrated correlation between endogenous testosterone and LBM [lean body mass]" and because "exogenous testosterone and endogenous testosterone operate differently within the 'extraordinarily complex' endocrinological system." Another expert witness, psychologist Sari van Anders, whose studies of T's responses to social context we described in Chapter 6, gave the example of interaction between T levels and receptors as one important element of that complex system. Van Anders explained that "three factors influence the effect of endogenous testosterone on a person's body, namely (a) testosterone levels; (b) androgen receptor function; and

(c) androgen receptor count/location." Recall that androgen receptor function is variable both across locations of the body and across different individuals. It is also dynamic: various factors can up- or downregulate it. Van Anders also explained that "exogenous testosterone sometimes has the opposite effect of endogenous testosterone."[49]

CAS rejected arguments about the difference between exogenous and endogenous T, seemingly swayed by the IAAF's scientific experts, who argued that there is no biochemical distinction between exogenous and endogenous T, and "once exogenous T is in the bloodstream, the molecule is the same and the androgen receptors cannot distinguish between exogenous and endogenous testosterone." But Chand's experts weren't arguing about the biochemical identity of endogenous versus exogenous T. They were arguing about the relationships and processes that are engaged by one versus the other. Exogenous T interrupts a system that is self-regulating. Recall the positive and negative feedback loops for T, which are better understood in men but likely also operate in women. At lower levels, T stimulates the hypothalamus and pituitary to release hormones that ultimately result in more T production. At higher levels, T suppresses the hypothalamus and pituitary, and T production drops lower. Van Anders cited evidence that further complicates these feedback loops—namely, that the sudden interference with T levels that happens when exogenous T is introduced also has effects on the activity of androgen receptors. She cited work by Sader and colleagues, who set out to investigate the mechanism underlying the commonly observed pattern that androgen receptor activity is generally higher in males than in females. Sader and colleagues concluded that it was a hormonal rather than genetic process, but along the way they made a fascinating discovery about the difference between endogenous and exogenous T exposure: "Exogenous hormone treatment, regardless of gender, sex steroid type (androgen or oestrogen) or intent (physiologic replacement therapy or pharmacologic treatment) differs from the effects of endogenous androgen exposure." There is no biochemical difference between T that is synthesized inside versus outside one's body. But interrupting the person's usual hormonal milieu by introducing T from the outside created a reaction. In trans men who took T, for example, and suddenly had much higher T levels than they previously had, their androgen receptors rapidly downregulated, meaning they were less active, effectively getting less bang for the buck from the T in their system.[50]

Opportunistic Epistemology

We opened this chapter with a story about divergent accounts of T and the role it plays in athletic performance. At the time we encountered these particular stories, we had already argued that the available science on T and women's athletic performance couldn't support the regulation. But hearing different experts advance such flatly contradictory claims none-theless made us wonder whether their differing accounts and disagree-ments were in part about multiplicity, stemming from their different dis-ciplinary approaches. The scientific experts that the IOC and IAAF initially gathered to construct the rule were not scientists who study sports and T, but doctors who work with people with intersex variations. This could explain the fundamental framing of women with very high T levels as out-side the gender binary and the regulations as necessary to restore that binary. Doctors in this field have long followed a clinical protocol that medicalizes and pathologizes atypical sex traits, imposing binary gender regardless of medical need. It's not surprising, then, that the regulations target women whose bodies don't conform to normative gender binaries, position those women as outside the group of women athletes who de-serve fairness, and amplify widespread prejudices about difference rather than addressing any demonstrated problem in women's sports.[51]

Seeing that the IAAF and IOC had tasked clinicians instead of sports scientists with crafting the regulations, we argued that they had called on the wrong experts, and thus put the wrong scientific "facts" into play. Science studies scholars have been showing for decades that the facts emerging from any scientific endeavor are messy, equivocal, and context-specific; the science of T is no exception, and neither is the science of T and athletic performance. As we have shown, the closer you look, the harder it is to make firm, global pronouncements. Having looked into the literature on T and athleticism, we saw the gaps and differences across contexts. But the regulations frame the facts on T and women athletes as unidimensional, singular, and conclusive. We initially considered this di-vergence in light of Longino's model of pragmatic epistemology, which aims to move beyond the appreciation of multiplicity (significant as that is) to understand the stakes embedded in divergent accounts of the "same" phenomenon.[52]

But we discovered that multiplicity alone can't explain the different versions of T and athleticism discussed in relation to the T regulations. IAAF policymakers and other proponents of the regulations opportunistically swing between scientifically appropriate caution about what data on T and athletic performance show, on one hand, and flat declarations about T's deciding role, on the other. If you read the IAAF testimony in the court's decision on the Chand case, for example, there is no whiff of scientific uncertainty or limitations to the evidence: T is the main ingredient in athletic success, and variability among athletes is reduced to the simple dichotomy between those whose tissues respond to T and those who are insensitive. Yet in the same year the case was heard, the IAAF published a paper noting that "the lack of definitive research linking female hyperandrogenism and sporting performance is problematic and represents another central point of the controversy. With the exception of data extracted from doping programs in female athletes in the former German Democratic Republic, there is no clear scientific evidence proving that a high level of T is a significant determinant of performance in female sports." There were earlier signs of equivocation. In a 2013 paper, the IAAF simultaneously claimed that women with high T have "a massive androgenic advantage" in athletic performance and demurred: "The male advantage in certain sports is *most likely* explained by the fact that men produce 'much higher levels of androgenic hormones'" and "All these effects of testosterone *could be* beneficial for physical performance." These equivocations hew closer to the available science, but they weren't going to win the IAAF's case in court.[53]

In arguing for the T regulations, the IAAF practices opportunistic epistemology—reverse engineering that starts with a course of action, a policy, or a conclusion and searches for evidence to support it. Instead of a "judicial" approach to data, this is a "lawyerly" approach, with the regulation as the client. In the case of the IAAF, this has meant crafting its studies with the express purpose of supporting its foregone conclusion that women with naturally high T should not compete with other women. But in order to appear scientific, the IAAF sometimes has to hedge, because the gaps and tentativeness are obvious to anyone who scratches the surface of relevant studies. So policymakers work the evidence and interpretive statements with two competing goals: establishing and maintaining their status as "scientific," and producing firm statements that seem to support a blanket regulation of the sort they had already implemented.

To dispel any lingering doubts that this is opportunism, consider the following sequence of events. In 2015, when CAS granted Dutee Chand's challenge to the regulation, it gave the IAAF two years to come back to the court with evidence that high-T women have a performance "advantage" over low-T women that is of roughly the same magnitude as that of elite men over elite women, or roughly 10–12 percent. When its window of opportunity to respond had nearly closed, the IAAF released its own study of the relationship between T and performance in women based on data from the 2011 and 2013 IAAF World Athletic Championships. This study had several problems. First, the version of T that was to be subject to the regulations (total T) was not related to any outcomes in the study. Second, the study found that a different measure of T (free T) was statistically significantly related to better performance in just five of twenty-one track and field events, with the top group performing 1.78 percent to 4.53 percent better than those with the lowest free T—far from the level of difference that CAS said needed to be demonstrated. For most events, T made no difference. In the original study, women in the lowest free T group actually did better than those in the top group for seven of eleven running events, although most differences didn't reach the level of statistical significance.[54]

Still, the IAAF heavily promoted the study, issuing a press release that heralded it as "new support" for the regulation. Criticisms emerged and quickly piled up, with a group of experts eventually calling for the study's retraction.[55] There have been multiple corrections and reanalyses published, all with different results. Across all analyses, women with lower T do better in some events, and in none of them do women with higher T levels do better in all the regulated events.

Meanwhile, the IAAF avoided answering these criticisms by pivoting away from its original T regulation, unveiling a reworked version in late April 2018. Titled "Eligibility Regulations for the Female Classification (Athletes with Difference of Sex Development)," the new version still draws on the idea that women with high T have a performance advantage, but regulates only long sprint and middle distance races (between 400 meters and the mile). Moreover, the T threshold is lowered to half the previous level. This could conceivably capture a much greater number of women, including those with the most common cause of naturally high T, polycystic ovarian syndrome (PCOS), which affects up to 20 percent of women. But as the title of the new regulation indicates, women with PCOS are explicitly excluded, putting into stark relief its true target: women with high T in the context of specific intersex variations. Perhaps the most blatant opportunism is evidenced by the fact that the events targeted under the new rule are different from those events in which the IAAF's own study found T to play a role; for example, its study found T related to hammer throw distance, but it's not a regulated event. It does, however, include the middle-distance running events in which athletes long targeted by the IAAF compete.[56]

Not long after the release of these regulations, South African middle-distance runner Caster Semenya and Athletics South Africa brought a case against the IAAF arguing that they constitute unfair discrimination and lack a sound scientific basis. They further argued that the regulations are "unnecessary to ensure fair competition within the female classification; and are likely to cause grave, unjustified and irreparable harm to affected female athletes." In a two-to-one split decision, the CAS arbitrators sided with the IAAF.

The decision reignited public controversy over the regulations, and supporters doubled down on their insistence that "science" is on their side. But they often make basic logical errors in evaluating relevant evidence. P. J. Vazel, an elite track and field coach and member of the Association of Track and Field Statisticians, notes the confusion between "*intra*-individual analysis where raising or lowering T will show a relationship with performance, and *inter*-individual analysis where there is no relationship found with performance." That is, a woman might perform better with her natural level and perform more poorly if her T levels are artificially lowered, but that isn't grounds for concluding that T levels are responsible for performance differences between women athletes. Among other possibilities, the drop in performance might be attributable to the many side effects and physiological changes that go along with dramatically lowering T. The illogic of inferring inter-individual difference from intra-individual variation is further established by the IAAF's own analysis of data from women at World Championship events.

Likewise, when confronted with the lack of evidence that T is related to inter-individual differences in performance, supporters of the regulations often fall back on the female-male difference in athletic performance. But you can't prove a relationship between T and performance based on female-male comparisons, because there are too many differences between men and women athletes at the group level. Still, many commentators, including the science journalist Gina Kolata and conservative political columnist Andrew Sullivan, drew on arguments put forth by Doriane Lambelet Coleman, a professor of law and a former elite 800 meter runner who vigorously supports the T regulations. Coleman claims that "the primary reason for the sex differences in the physical attributes that contribute to elite athletic performance is exposure to much higher levels of testosterone during male pubertal growth. Those physical attributes include power generation, aerobic power, body composition and fuel utilization. Compared to females, males have greater lean body mass (more skeletal muscle and less fat), larger hearts (both in absolute terms and scaled to lean body mass), higher cardiac outputs, larger hemoglobin mass, larger VO_2max (i.e. a person's ability to take in oxygen), greater glycogen utilization, and higher anaerobic capacity." That there are sex differences in a wide range of physical attributes that contribute to athletic performance is uncontroversial. But attributing all those differences to T and, further, saying that circulating T level drives and maintains all those features, is not a consensus opinion among scientists. Moreover, it is precisely because there are so many other differences between women and men athletes, both physiological and social, that many scientists don't consider male-female comparisons a useful form of evidence for understanding how T affects athletic performances. Female-male comparisons are too confounded, so only within-sex analyses can give clear enough information about the specific role of T. Yet Coleman has claimed that arguments against the assertion that T is well established as the main ingredient in athletic performance "have gotten no traction except in circles where the science and its implications for medicine and doping are not well understood." The surface of "the science" of T in Coleman's version is slick, eschewing complexity and gaps in knowledge. Hidden beneath the smooth surface is the fact that Coleman draws only on the same evidence that IAAF presents.[57]

In their 2019 decision, the CAS arbitrators expressed several "serious concerns." One was about the side effects of hormonal treatment and the "practical impossibility of compliance," perhaps because manipulating T to a specific level is not straightforward. The second was the insufficient evidence that higher T levels produce "significant athletic advantage" in the 1,500-meter and the mile races. Nevertheless, they left it up to the IAAF whether to regulate those events. When IAAF president Seb Coe was asked whether they would remove these races from the regulations, he simply replied, "No."[58]

T as the Great Distraction

In 2013, doctors affiliated with the IAAF published a report illuminating what happens when women are investigated under the T regulations. Four young women, aged eighteen through twenty-one and from "rural and mountainous regions of developing countries," were identified through various means as having high T, and each was sent to the IAAF-approved specialist reference center in Marseilles, France. A multidisciplinary team of clinicians conducted extensive investigations aimed at assessing sex-linked biology, beginning with endocrine, karyotype, and genetic analyses. They also inspected the women's breasts, genitals, body hair patterns, internal reproductive organs, and basic body morphology in detail, and interviewed the women about their gender identity, behavior, and sexuality. The doctors pronounced, based on the wide-ranging physical and psychological exams, that the women's T was "functional," signaling that they don't meet the clinical criteria for complete androgen insensitivity, which would have exempted them from a T threshold under the regulation. They also determined that the women's high T levels could be explained by intersex variations, specifically chromosomal variations and internal testes. Although the doctors acknowledged that leaving the women's testes intact "carries no health risk," they recommended that they be removed, on the grounds that gonadectomy would "allow them to continue elite sport in the female category." But the medical team aimed for more than lowering T. The doctors also performed surgical and medical interventions long practiced with the aim of "normalizing" girls with intersex variations, including "a partial clitoridectomy . . . followed by a deferred feminizing vaginoplasty and estrogen replacement therapy."[59]

The paper violates the athletes' privacy and confidentiality and should not have been published. It sheds light, however, on an implementation process that is otherwise kept under wraps, and further highlights whom this regulation burdens. The genital surgeries described in the report suggest that something beyond T and athletic performance motivates the regulation, and indicate that it is not just compliance with the T regulation that drives the interventions. Sports authorities, through public talks, publications, and interviews, have consistently indicated that the women investigated for high levels of naturally occurring T are exclusively from the Global South, and all indications are that they are black and brown women.

Two of the most fundamental narratives in T folklore come together to undergird this extraordinary and irrational violation of the four young athletes: T as "jet fuel" for athletes, and T as the "male sex hormone," out of place and dangerous in women's bodies. T talk is fork-tongued: not only does high T supposedly provide an "unfair advantage" to women athletes, but it also makes them sick. Policymakers have justified the policies as not simply a matter of fair competition but as necessary because naturally high T in women is a health problem, such that sports authorities have "a duty within the context of medical ethics" to identify women with high T and direct them into treatment "to protect the health of the athlete." The health justification is embedded in the regulation texts: the IAAF claims in one version that the regulation is for "the early prevention of problems associated with hyperandrogenism," with a 2011 IOC press release reading, "In order to protect the health of the athlete, sports authorities should have the responsibility to make sure that any case of female hyperandrogenism that arises under their jurisdiction receives adequate medical follow-up." High T, though, does not constitute a medical problem, and physicians do not lower T in the absence of patient complaints or functional impairments. Lowering T can cause significant health problems, which can include depression, fatigue, osteoporosis, muscle weakness, low libido, and metabolic problems; these effects may be lifelong, and may require hormone replacement treatments, which are both costly and often difficult to calibrate. Some of the interventions will leave athletes sterile.[60]

Ignoring the problems associated with dramatically lowering T and unnecessary surgeries, the IOC and IAAF frame interventions as an unmitigated good, especially because they target women in regions from the Global South that policymakers have described as "lacking competence" for medically managing intersex variations. Yet the interventions on athletes are directed not by the athletes' goals and needs but by the goals of sports organizations. Neither the regulations nor any sports officials' publications or presentations that we have encountered acknowledge the now decades-old controversies that have raged over genital surgeries and other medical interventions for intersex. The interventions performed on women in order to comply with the regulation are the same ones that adults with intersex variations have argued against for decades, pointing out that they are driven by gender ideologies that pathologize sex-atypical bodies and behavior that breaks gender norms, and that these interventions cause irreparable harm to sexual sensation and function. Influenced by these arguments, delivered forcefully from individuals in countries around the world, national legislative bodies and human rights organizations have responded with condemnation and in some cases legal prohibitions against these practices.[61]

Sports officials move between two justifications for the regulation—protecting health and protecting fairness—but for the women being "protected," these two arguments are mutually exclusive. Women with high T are not "visible" in the fairness portion of this regulation except as a threat; the "help" offered requires that they submit to pathologization despite having no health complaints.

The regulation rests on the premise that "women athletes" are a vulnerable class that needs protection. But from whom? History is full of examples of how the "female vulnerability" argument has benefited women with more privilege (whether from class, race, sexuality, gender presentation, or region) over women with less privilege, who are ironically but systematically seen as less vulnerable. In a context in which T alone is deemed to determine advantage and disadvantage, what makes sense and is valued as legitimate is the need to protect women athletes with lower T from presumably "unfair" competition. But women investigated for possible high T face harms that are nowhere in the calculus: having their identity publicly debated, their genitals scrutinized, the most private details of their lives assessed for masculinity, and their careers and livelihoods threatened, and being subject to pressure for medically unnecessary interventions with lifelong consequences. The narrative of harm is inverted: how does the putative advantage conferred by T matter more than concrete and demonstrable harms to people?[62]

T talk deflects attention from the racial and regional politics of intrasex competition in women's sport, obscuring how the regulation systematically and materially harms those with less power and privilege. In an alchemy between folklore and science, T talk validates familiar cultural beliefs as scientific, and scientific accounts get a free pass on some of the details and consistency that they would otherwise need to provide. As with the idea that dominance, power, and aggression are about T, T talk lends "truthiness" to the ideas that T is both male and the main ingredient in athleticism, thus certifying the regulation as rational. And as in those domains, T talk deflects attention from social structures and institutions, attributing the result of competitions completely to individual bodies, as though these bodies have developed, trained, and ultimately competed in some socially neutral vacuum.

Ironically, T's authorized biography, with its pat story about how T fuels male-typical athletic performance, is a powerful distraction from T itself, occluding the fascinating, diverse, and contingent actions of T. This hormone doesn't drive a single path to athletic performance, nor even a small set of processes that can be linearly traced from more T to more ability. T is involved in many of the processes that underlie athletic performance for most people, but it should come as no surprise that it's neither a sufficient nor even necessary ingredient. The classic example to consider here is women with complete androgen insensitivity syndrome, who are overrepresented among elite women athletes, though studies show they have no ability to respond to T at the cellular level. Just as T isn't simple, neither is athleticism separate from other human capacities: athletes must develop these capacities to a very high level, but at lower levels, strength, flexibility, coordination, and motivation are required for basic survival. It wouldn't make any sense for us to have evolved in a way that put that range of essential capacities under the control of any singular variable, even T. [63]

CONCLUSION

THE SOCIAL MOLECULE

T's AUTHORIZED BIOGRAPHY—the unifying story of a substance that carries out a process of sex differentiation across multiple scales of space and time—is both carrying on and being challenged from multiple directions. In scientific work, the authorized biography is still an important conceptual reference point that shapes the questions asked, the people chosen as research subjects, and scientists' ultimate interpretations of their findings as they fit their own studies into the larger body of knowledge about T. At the same time, many people are thinking about T in different and interesting ways. Researchers routinely amend and challenge the classic tale with concrete data: T's actions can't be fit into a linear story about accumulation of masculinity. Scholarly work in diverse modes—from Emilia Sanabria's ethnographic study of hormone use to Paul Preciado's queer auto-ethnographic manifesto on T to Sari van Anders's social neuroendocrinology studies—are using T in projects that aim to reconsider, disrupt, and reshape gender. But these interventions don't replace the classic tale: T talk manages to endure, overshadowing or coexisting alongside whatever new narratives emerge expressly because it is diffuse and flexible.[1]

Throughout the book, we've drawn on concepts from science and technology studies (STS) to help make sense of how scientific practice around T unfolds, and especially to understand the inextricability of culture and nature in the facts that emerge from scientific research, circulate and morph in the world, are incorporated again into science, and so on. We are fascinated by, and think it is important to understand, how facts about T are constructed, especially if we are to appreciate the work they do in the world. Readers who are most familiar with STS will have already seen, though, that we also work in a vein that is less typical for that scholarship. This is largely because in the course of examining how T facts are constructed, we have often challenged scientific practices and the claims that circulate through T sciences. Empirical challenges and critiques are not usually the point of STS, and can sometimes distract from the larger aim of STS studies. As the physicist and STS scholar Karen Barad has so beautifully put it, "Empirical adequacy is not an argument that can be used to silence charges of constructivism. The fact that scientific knowledge is constructed does not imply that science doesn't 'work,' and the fact that science 'works' does not mean that we have discovered human-independent facts about nature. (Of course, the fact that empirical adequacy is not proof of realism is not the endpoint, but the starting point, for constructivists, who must explain how it is that such constructions work—an obligation that seems all the more urgent in the face of increasingly compelling evidence that the social practice of science is conceptually, methodologically, and epistemologically allied along particular axes of power.)"[2]

Empirical adequacy is, of course, a matter of judgment and negotiation. The fact that we have labeled our biography "unauthorized" signals from the start that we take issue with the empirical adequacy of some very common accounts of T. Going beyond or perhaps behind the question of empirical adequacy, there are broader lessons about the way that scientific facts about T are constructed that can be seen from the vantage point of looking across the various chapters, and we use this Conclusion to summarize those. First, T is doing much more than the masculinization frame allows. T's diffuse effects are all the harder to model because of the persistence of the sex hormone concept. Second, T is a superb example of Barad's concept of intra-action: instead of discrete entities existing independently and then acting with each other (as implied by "interaction"), entities or phenomena emerge through engagement; they "make each other up" in an unbounded, fluid process of mutual influence. T emerges through the intra-action of multiple aspects of nature and culture. Because the specific material properties of T include a capacity to respond to social situations, the intra-active nature of this molecule can help to extend some important STS concepts, as we explain shortly. Third, some contemporary accounts of pharmaceutical T suggest that people think of T as a precision technology: using T to achieve specific, predictable, and sometimes quite narrow outcomes. While we have mostly explored research on endogenous T, that work challenges the idea that T as a technology can deliver what many seem to want from it. Fourth, while pre-theoretical assumptions about gender have long been recognized as shaping research on T and other hormones, scientific engagements with T also mobilize long-standing ideas about hierarchies of race and class. In important ways, this makes both race and class distinctions seem as though they emanate from biology rather than social dynamics and structures. We close the chapter with some final thoughts about how scientific authority regarding this molecule is shifted by the converging consensus that T is an entangled product of nature and culture.

Outside the Frame of Masculinity

While most research on T continues to treat the hormone as the essence of masculinity, there are plenty of researchers who approach T as a more multipurpose molecule. Escaping the masculinity frame radically changes the interpretation of T's capacities. Think of Blair Crewther's studies on interventions with athletes, where he and his colleagues see relationships between T and athletic performance only in certain groups of athletes, under certain circumstances, and along certain parameters of performance, in an overall pattern that can't be explained with a simple "masculinization" frame. Instead of suggesting that T's effects are "weak" or that T isn't doing anything important where they don't find relationships, Crewther thinks the better explanation is that T has diverse effects across multiple systems, and that those effects may end up canceling each other out if the researcher is looking at a composite variable like performance on a specific athletic task.

Crewther and colleagues' study of T and weightlifting performance in young Olympians is especially clarifying. Given that T tends to have positive effects on muscle mass and strength when paired with intensive weight training, this might be a subgroup of athletes where the usual mixed or null relationships would not apply: it seems logical that T would be related to weightlifting performance for this group. But it was not. As a potential explanation, Crewther focuses on the fact that T affects both muscle and fat tissue. Moreover, effects vary both across individuals and across regions of the body (e.g., lower legs and shoulder areas have different patterns of tissue response to T). While those inter-individual and inter-regional differences in effects sometimes work together to improve performance, sometimes the overall effect is actually negative, and other times it's simply a wash.[3]

Step back for a moment to consider how this looks different from a more traditional approach that simply evaluates T's effect on "athletic performance" or on "strength," both of which are either implicitly or explicitly coded as masculine. Sometimes researchers seeking correlations between T and athletic outcomes don't find the connection they have hypothesized. When this happens, they may check for or propose mediating variables: either those that might obscure or "interfere" with their ability to perceive T's effects, or those that directly affect performance in a way that opposes the generally positive effect they assume T will have. Crewther and colleagues suggest something else: T's effects aren't one part of the puzzle to be slotted in with other actors, but are themselves flowing through diffuse systems in ways that aren't predictably related to a composite outcome like "performance." While Crewther never put it this way in his publications or our conversations with him, his approach unburdens T from always having to carry masculinity, and through that new frame, it's possible to perceive what T may actually be doing.

Another clear example of where T's actions can't be understood as "masculinizing" is in female reproduction and the early stages of follicle maturation. As we write, acceptance of the idea that there is an optimal level of androgens necessary for follicular development is becoming more widely accepted. While giving T directly is not generally the treatment of choice (for reasons we touch on in the "T as a Precision Technology" section in this chapter), a consensus is developing that T, and not some other steroid like DHEA or estrogen, plays an obligatory role in recruiting the early-stage or "primary" follicles to enter that cohort of follicles that mature in later stages. This is yet another action in T's portfolio that could (and we think should) cause people to rethink the entire idea of classifying it as an "androgen." Even Norbert Gleicher, who was concerned about making comments that might seem "too PC," allowed that the sex hormone concept might be part of the reason that this knowledge about T's role in ovulation has taken so long to emerge.[4]

While the examples we've just provided give a more capacious view of T's effects by focusing on bodily tissues and physical capacities, Sari van Anders has proposed a model that rethinks the social and emotional processes T participates in. Against the pre-theory frame that predicts T will promote masculinity, van Anders suggests that T is involved in trade-offs between competition and nurturance, social aims that are relevant to both females and males across species, rather than specifically gendered. What each of these examples has in common is an ability to make sense of previously contradictory data. Together, they reveal a T that's fundamentally at odds with the authorized version of the "male sex hormone." T is not, at root, evolution's proximate mechanism for generating either masculinity or heteronormative coupling. It's a transcendent, multipurpose hormone that has been adapted for a huge array of uses in virtually all bodies.[5]

Amplified Intra-Actions

We opened this book with an episode from *This American Life* that explored, in the host's words, "testosterone and just how much it determines of our fates and our personalities." A better understanding of T might follow if the question was inverted, asking how much our fates and personalities determine our T (though even here, we'd be on firmer ground if we talked about influences and shaping rather than determinants). Unlike its authorized counterpart, this unauthorized biography foregrounds T's relationality, and emphasizes that part of the relationship between T and the social world that is most consistently supported by research, namely that T reacts to social situations.

To understand the implications of T's fundamental responsiveness, we draw on two foundational concepts in STS: Barad's intra-action, described earlier, and Donna Haraway's "natureculture." Natureculture is a WYSIWYG concept that signifies the inseparability of nature and culture, their mutual and relational constitution: "Flesh and signifier, bodies and words, stories and worlds: these are joined in naturecultures." Haraway situated the concept of natureculture in the era of technoscience, but our voyage in the world of T research suggests that fused phenomena of natureculture don't require the operations and innovations that inhere in a technoscientific world: T is natureculture at its core.[6]

T's biocultural entanglements operate at multiple levels. The first has to do with how we know or study T: while it is a specific material molecule, like other aspects of nature and material reality, we have no way of understanding it except through our human engagements, linguistic forms, cognitive capacities, and particular scientific tools. The second has to do with T's "being" or ontology. T recursively participates in processes that connect human social situations, perceptions, and emotions with biochemistry. T's specific capacities derive from its relation to other actors in its orbit: other steroids in the bloodstream, proteins to which it might be bound, the presence and properties of receptors, and so on. Because these actors are affected by aspects of the social and material world, including at a minimum nutrient intake, exercise, and physical environment, T is likewise implicated in those relationships. Without trying to elaborate every possible enmeshment that constitutes T as natureculture, it is already clear that T is a natureculture phenomenon whether humans are looking at it or not. While Haraway elaborated the theory of natureculture to describe the conditions of existence within the specific historical and political economic situation of late capitalism and an era of biotechnology, T seems to be a natureculture phenomenon that does not require reference to a specific form of political economy or technoscience. T is not special in this regard; work in social epidemiology and in STS increasingly suggests that natureculture, as an ecological frame, is a good general description of organic entities.[7]

What has been less elaborated is the way that the fundamental responsivity of T intensifies the meaning of intra-action. Critical analyses of hormone research and concepts constitute an especially rich subfield of STS, but so far, the insights in those studies neglect or underplay the specific material capacities of T and other hormones to respond to social situations. Hormones are not seen as "stable objects" in this literature; rather, their mobility or fluidity is theorized with attention to pharmacological intervention and the processes of steroidogenesis, rather than attending to the meaning of responsiveness as a feature of these molecules. Thinking about hormones connects the body with the social and emotional world in a similar way to the way Elizabeth Wilson connected the gut to "minded states" in *Gut Feminism*: "not that the gut *contributes* to minded states, but that the gut is an organ of mind." In the case of T, both scientists and the lay public already treat this molecule as a "chemical of the mind," but few appreciate its role as a chemical of social relations.[8]

Scientific and philosophical accounts of hormones converge on the concept of relationality. This rare alliance is achieved by coming at hormones from different directions. STS and philosophical accounts of hormones have established that they are social, relational objects by studying scientific practices and the way that research findings incorporate scientists' social worlds, cultural formations, and apparatuses. Scientific studies arrive at a relational, social account of hormones by examining how they behave in different social circumstances, especially observing that experimentally manipulating social circumstances prompts the endogenous production or suppression of specific hormones. This brief summary indicates that there is not an exact alignment between what STS scholars and hormone researchers mean by hormones being social and relational, and suggests that the STS version of this insight does not typically go far enough.

Research we have described suggests that the exercises we do and don't do, the family situations and roles we take on, our social stresses and feelings of frustration or mastery all participate in our bodies' production of T. In other words—and as a Baradian view would have anticipated—T does not exist prior to the intra-actions that (re)constitute its entanglements. T is literally manufactured in the body under specific circumstances that never exactly repeat: external and internal events that we still only dimly perceive set off the different steps of steroidogenesis. T is not simply in the body waiting to facilitate changes in tissue or spark neurotransmissions that tip the scales in favor of one behavior or emotion over another. T is called forth and comes into being at the very moment of those engagements. But there is more: these micro-level aspects of social context are themselves embedded within macro-social formations. If there is a critical takeaway from earlier STS studies of the so-called sex hormones, it is the way that knowledge about these hormones has been fundamentally shaped through gender ideologies, best demonstrated by Nelly Oudshoorn's influential exploration of "the invention of sex hormones." Drawing on more recent hormone research, we find another direct line between gendered formations and T: not through scientific practices, but through the hormone's ontology as a fundamentally social molecule.

Historians, critical biologists, and others who have conducted STS analyses on hormones have done groundbreaking work on scientific practices relating to hormones. Oudshoorn's description of the way physical assays were chosen to signal the "essence" of masculinity or femininity prior to isolation of specific hormones is a beautiful example of the way that practices materialize certain objects: androgens and estrogens were brought into being through those measures. We want to call attention to the way that contemporary scientists' "apparatuses" (think: schemes for recruiting subjects, questionnaires, hypotheses, laboratory games, saliva

sampling kits, statistical tests, and more) embed T in particular social relations while excluding others. As Barad insists, scientific "apparatuses
must be tuned to the particularities of the entanglements at hand. The key
question in each case is this: how to responsibly explore entanglements
and the differences they make." Specific apparatuses employed to demonstrate T's relationality and responsivity have overwhelmingly materialized
(or, to use Annemarie Mol's term, "enacted") T in relation to emotions and
social contexts that have a masculine valence: competition, power, dominance, and aggression. Answering Barad's call to a responsible exploration of the entanglements of research practices with T will require, at a
minimum, acknowledging and breaking the habit of forcing T into these
narrow associations.[9]

T as a Precision Technology?

In the book *Testo Junkie,* the philosopher Paul Preciado describes a twelve-
month experiment of taking T daily as an "auto–guinea pig," aiming to
conduct a "do-it-yourself bioterrorism of gender." Using the model of computer hackers' "copyleft" programs, which put information technology
and its transformation into free circulation as a political act, Preciado describes a potential "gendercopyleft revolution" that could reconfigure
gender itself by distributing hormones outside the closed circuits of com-
mercial production and state regulation. With Testogel sachets from the
black market, Preciado's aim was not to transition, but rather to explode
gender.[10]

   Preciado's book is part of a burgeoning literature and popular conversation about hormones that, in varying degrees and with radically different
strategies, demonstrate that people of many different stripes use T to reconfigure or disrupt gender, and not just at the level of the individual. Used
politically to shift gender and power structures, we might think of hormones as a tool of bioanarchism: refusing the regime of gender, hormones
and hormonal (self) knowledge could be distributed in novel ways that
subvert the idealized package of "bodyminds" that are perfectly male or
perfectly female. Insofar as Testogel and other T products can help people
who aren't cisgender men to reshape their bodies as more masculine, T is
a useful tool for this project. When people use T to disrupt gender, they
seem to approach T as a "precision technology" that can selectively deliver aspects of masculinity just where it's desired. In *Plastic Bodies,* for

example, Emilia Sanabria follows cisgender women in Brazil who use pharmaceutical T, and finds that the circulation of hormones creates new potentials for "gender bending." Sanabria's project traces what people—both doctors and lay people—think they are doing with these hormones, how they think the hormones enable them to precisely retool gender. Hormones, especially estrogen and testosterone, are still "coded as female and male" by her informants, but "androgens (such as testosterone) [are] summoned with no apparent contradiction for patients or doctors in the making of new forms of femininity. With testosterone, women could be like men and yet remain women, we are told. They could become superwomen. Super-desiring and desirable. And perhaps, above all, super-productive." For some, it might seem obvious that T can't be used as a precision tool to confer vigor, focus, libido, and a sense of power. But our travels with T suggest that the majority of people would find it plausible and intriguing that you might use T in these ways. These "new projects" of using T walk a fine line between disruption and recapitulation of sex hormone ideology.[11]

Precisely because so many people find the idea of T as a precision technology credible, we think it's crucial to bring these practices into conversation with the materiality of T, and ask what, exactly, rejiggering T levels can accomplish. T and other hormones are technologies or prostheses insofar as they can be used for specific aims or goals. When hormones are used to control fertility, for instance, they could be understood as "prosthetic technologies," at least insofar as they are quite reliable in delivering on the promise of contraception. But a materialist has to be skeptical of the idea that people can use T as a behavioral prosthesis, because cumulative evidence on the material capacities of T suggests that it doesn't have any direct effects on people's behavior. T is malleable, but our ability to deliberately mold our lives and the world through T is limited. We can mobilize the material capacities of T to effect bodily changes that signal masculinity, especially by introducing amounts of exogenous T that are significantly higher than what a given body is accustomed to. We can also possibly mobilize social situations to stimulate or depress endogenous T. But it's very unlikely that T can be used to directly increase libido; decades of pharmaceutical experimentation and marketing have conclusively demonstrated that this just isn't an effective way to boost sex drive in men or in women.

Consider this alongside Preciado's experimentation with T. Even as Preciado delivers an astute account of the history of development of T as a "sex hormone," his experience with T and the theory it generates adds to the lore of T as the "hormone of desire." In Preciado's hands, this gives rise to sophisticated reflections on how T, in the hands of pharmaceutical companies, becomes the ideal marketable commodity: a product that foments desire, where desire itself is promoted as a good that people are increasingly obliged to feel, and a method for generating more demand for both this product and consumption in general.

As compelling as this is, Preciado's own extended narratives of sexual experiences with and through taking T amplify the historical marketing of T as generating and sustaining peak desire. Preciado's account is data with a certain kind of authority, from a smart and self-aware first-person narrator. Preciado's description of his own experiences strongly resonates with the authorized biography of T, thereby increasing the truth value of that particular story. At the same time, it's impossible to read this account alongside the placebo-controlled studies, which show that increasing T in healthy people has little to no effect on libido. Researchers have, however, noted an especially strong placebo effect in studies of hormones and sexuality.

Paying more attention to how personal versus scientific accounts come to be, it's clear that T might produce desire in some people out in the real world even if it doesn't produce desire in the lab. Begin by noticing that the relationships between T and various emotional and cognitive states are very different in controlled versus uncontrolled research. In uncontrolled or "open label" studies where people know that they are getting T, researchers have observed relationships between T and aspects of mood, cognition, and sexuality, including libido, in healthy people. But in placebo-controlled studies there is little to no relationship between T and any of these domains at any dose. It may be the case that T's effects on libido are only evident at really high doses, when someone's T levels are increased well beyond their typical individual levels, which is suggested by one meta-analysis on data in men. The most rigorous test of the relationship so far is probably Shalender Bhasin's randomized clinical trial of sixty-one healthy men, aged eighteen to thirty-five, that we discussed in Chapter 7. Even though men received supraphysiological doses of T in amounts that went up to 600 mg (twelve times the dose Preciado reported using), "sexual function, visual-spatial cognition and mood . . . did not change significantly at any dose" of T. It's not to say that T is completely unrelated to sexual function, including libido, but to note that the relationships are complicated and limited. In one study in which Bhasin and colleagues suppressed the endogenous T of healthy older men (aged sixty to seventy-five) and assigned the men to receive varying doses of pharmaceutical T, some aspects of sexual function did improve with T, especially at higher doses—but only for the men who were sexually active to begin with. Even studies that show associations between T and sexual function indicate that the effects are extremely modest, and so far they have only been demonstrated in either older men or men with low T and/or sexual dysfunction. Remarkably, only two tiny placebo-controlled clinical trials examine the effects of T on libido or other aspects of sexual function in healthy women: one shows an effect, and one doesn't. Large cross-sectional studies have not found any significant link between T and sexual function in women, though T therapy does show a statistically significant but very modest effect on libido in women diagnosed with the "disorder" of low sexual desire.[12]

At the same time, what Preciado reports is in line with the way many people who have taken T say it has affected their libido. Controlled studies can't capture a situation like Preciado's: it's not possible to do a placebo-controlled study of the effects of high doses of T in someone whose body has developed with and is habituated to much lower levels. The physiological effects of T on musculature, hair, and other observable features would make it obvious who is getting placebo. Some evidence suggests that trans men experience an increase in sexual function after transition, but this is unsurprising: trans people are forced to narrate sexual dissatisfaction in order to meet clinical criteria for medical interventions. It's also the case that many people take T precisely because they want and anticipate that it will increase their sexual drive and performance: taking T is an aspirational act. There's no reason that T should be exempt from the placebo effect.

But we think there's something more going on. T's diffuse effects include observable bodily changes, responses to situations and emotions, and more, creating loops of influence that connect the surface of the body to social worlds to interior physiological processes to emotional states and so on. In his memoir *Amateur*, Thomas Page McBee describes a loop that travels from taking T as a trans man, to the profound changes that are visible on his body's surface, to the social world of gendered interactions. "As the testosterone took hold and reshaped my body," he wrote, "its impact as an object in space grew increasingly bewildering: the expectation that I not be afraid juxtaposed against the fear I inspired in a woman, alone on a dark street; the silencing effect of my voice in a meeting; the unearned presumption of my competence; my power; my potential." In McBee's telling, the T doesn't change his behavior in a way that ripples out from his brain to the world. Instead, T forces a change in his behavior because his body is now a different social stimulus. People see and respond to a man, and this new situation requires and rewards different behaviors.[13]

We can imagine similar loops that are more internal to a person taking T. A thought experiment could trace the route of T's effects from changes on the surface of the body, like more facial hair, oilier skin, a receding hairline—signals that are all strongly coded as masculine and inform us "something profound has changed here." Recall Griffin Hansbury, whose observations after taking T were chronicled on *This American Life*. As we described in the Introduction, Hansbury experienced enormous changes, including in his attention and libido. When we sat down with him later, he compared those experiences to a young man's passing through puberty: the tracking and noting of every feeling, behavior, and physical change, but also the curiosity and eagerness to see how these changes shaped how he was seen and how people interacted with him.

Whether T increases in a typical puberty or later in life because a person deliberately raises their T, the bodily changes they experience may very well call forth other parts of T's assumed portfolio, especially those effects that are most often repeated in our cultural stories about T: libido, aggression, physical strength, and confidence might be at the top of this list. For this reason, it is pretty much impossible to do placebo-controlled research on T at high levels in people whose bodies are habituated to low levels of T: T's effects in this situation are diverse and not entirely predictable, but generally include an increase in facial hair, changes in skin texture and voice, and a greater ability to build muscle bulk through exercise. In fact, T's capacity to stimulate effects at the body's surface is one of the reasons that direct treatment with T is not the usual therapeutic choice for women with low ovarian reserve for whom androgen stimulation could improve the number of eggs they produce for in-vitro fertilization. In these patients, even relatively low levels of T can stimulate effects that are often unwanted, whereas giving the upstream hormone DHEA stimulates naturally proportioned changes in the levels of various downstream steroid hormones.

As a technology, T hasn't been quite as satisfying as some other hormones, specifically estrogen and progesterone, that have been wrangled into a formulation that reliably prevents pregnancy. Although it's not possible to use hormones for contraception without also increasing risk for some cancers or stroke, especially for women who smoke, the stakes for getting pregnant are high enough for many women that they're willing to make those trade-offs. Hormonal contraception is by some measures a rough tool, but as medicine goes, it is relatively precise: women who use hormonal contraception according to the instructions can be confident that it dramatically reduces the rate of pregnancy. In contrast, even when people increase their T levels far beyond what their bodies have ever experienced, they will almost certainly get physical effects, but not necessarily the physical effects they're looking for. For instance, many trans men express frustration that T doesn't create much of a change in musculature unless the men also follow a serious exercise regimen, preferably weight training. This converges, coincidentally, with clinical trial descriptions of how T affects muscles, as we describe elsewhere in the book. Men also express a wide variety of experiences with how T affects facial and body hair, with some developing thick beards and others never getting past the wispy stage. In other words, the variation is much like that in cisgender men, and the degree of "masculine" body response doesn't map onto how much T the individual men are taking.[14]

Throughout the book we've looked at the way materiality and narrative assemble in scientific projects to make knowledge about T. Facts that emerge from personal experience are likewise collected and curated. As the historian Joan Scott has observed, personal experience is often given a greater truth-value than other forms of evidence, but personal experience is as much mediated by epistemic frames and historical and sociocultural specificities as is other evidence, such as clinical or psychological studies or sociological research. Everyone isn't drawing on exactly the same frameworks to understand and engage with T, but the core narratives of T talk are broadly disseminated and rarely challenged, so the same narratives of masculinization that shape scientific explorations of T also shape personal experience. This is as true of aging men with lower T who hope that supplementing with T will bring them closer to the masculinity of their youth as it is of the women in Brazil and elsewhere who are hoping that small doses of T will give them little bits of masculinity, à la carte, and of people who are more self-consciously trying to hack their gender. But both the mainstream and the revolutionary versions of metamorphosis via hormones rest on the belief that T conveys not just bodily masculinization, but masculine gender. This in turn contributes to a shared sense of what you can "do" with T, and an inflated confidence that T can be directed toward specific ends.[15]

Submerging the Social in Hormones

T is often a mechanism for individualizing the social, most obviously with gender. Too often, T is invoked as an apologia for a status quo sexism in which sexual violence, higher male salaries, men's overrepresentation in prestigious occupations, and the tasking of women with domestic drudgery must all be accepted as natural and inevitable. Scientific papers as well as popular accounts of T and aggression, for instance, often open with statistics on higher male prevalence of most forms of aggression, but seldom discuss how gendered institutions and gender socialization shape this disparity. The structure of gender essentialist claims is more subtle than it used to be, and may sometimes reflect the frustration and bewilderment of people who feel that more gender equality would be fair, but is just too difficult to achieve. Why haven't forty years of anti-discrimination legislation and educational policies equalized the number of women and men in STEM fields or finance? Where are the women entrepreneurs in Silicon Valley? Why is #MeToo overwhelmingly a movement of women detailing abuses by men? T provides an easy answer that, if not exactly comforting, nonetheless absolves us collectively from having done anything wrong. It's not "us"—it's T.

But the way T operates in discourses of power goes beyond gender. One of our most significant aims with this book has been to bring light to how T research makes race hormonal. There are few precedents for our exploration of the racialization of T. Evelynn Hammonds and Rebecca Herzig have described how "glandular differences" figured in accounts of race in earlier American biomedical sciences. Based especially on endocrine studies from the 1920s to the mid-twentieth century, they observed, "Just as 'genes' appear in most twenty-first century social controversies, so, too 'glands' once seemed to hold the promise of definitive answers to difficult social issues." By the early 1940s, the New York criminologist William Wolf adopted the term "endocrinopathy" to describe those motivated to crime by bad glands. Criminality, sexual depravity, susceptibility to disease, feminism, and labor agitation were all tied to "glandular derangement." Celia Roberts's critical explorations of the promotion of "hormone replacement therapy" demonstrate how research on menopause, combined with pharmaceutical product development and marketing, mobilized racialized views of non-Western women as closer to "nature." She also briefly points to a profoundly racialized model of evolved reproduction strategies in which T plays a prominent role. The *r/K* selection theory model, which we explore at length in Chapter 6, has been used by the psychologist Richard Lynn and others to propose a racial ranking of testosterone levels that would correspond to racialized patterns of "investment" in quantity versus quality of offspring.[16]

Most research on T is more subtle in its racial content. The models that researchers in behavioral endocrinology use to explore T and behavior have become much more complex in the past few decades, and these more sophisticated endocrine concepts obscure the fundamental attachment to a theory of "naturalized race." But research on T naturalizes race in at least three ways. The strategy that stretches across the most domains and specific studies is something we have come to think of as "sample magic," which is not observable in any individual studies, but only emerges as a pattern when you compare different studies that examine specific constructs within particular samples. Studies in the domains of risk, aggression, and power demonstrate this pattern. T researchers who study aggression or psychopathology draw their samples from specific domains: low-income, often racialized communities and settings like prisons and behavior intervention programs into which poor and marginalized people are funneled. By contrast, T researchers who examine the supposedly more "human" strategies for achieving dominance, such as subtle power moves in negotiation games, often draw their subjects from university populations. Likewise, researchers who study risk in marginalized people tend to characterize those risks as "deviant" or "antisocial," while those who study risks in MBA programs, Ivy League schools, business executives, and entrepreneurs frame risk-taking as essential to achieving dominance, and ultimately increasing fitness. Sample magic means that implicit assumptions driven by race (as well as class and gender) ideologies drive the selection of research subjects in particular domains. Usually without ever even mentioning race, such studies reinforce racial stereotypes in a manner that might be all the more powerful because it is so invisible.

A second way that T research naturalizes race is through resonance with widely circulating tropes of race, and linking commonly racialized behaviors with differences in T. In other words, even if a study doesn't examine racial differences, if a behavior or trait associated with racial stereotypes (such as aggression-as-black or leadership-as-white) is linked to T, then triangulation with offstage variables allows these studies to function as cognitive resources in the overall re-biologization of race.

The third strategy of naturalizing race as biological is specific to biosocial models, in which social structures are approached as essential features of racialized people. In an era of widespread agreement among social scientists that race is a social category, not a biological one, the sociologist Ruha Benjamin argues that "explicit racial references . . . are no longer practical." Benjamin envisions "race as a kind of technology . . . requiring routine maintenance and upgrade"; "innovators find ways to embed racism deep in to the operating system." This provides a way to understand how biosocial work on racial differences in T, which on the surface reads as if racial difference is the product of a body ingesting the conditions of racism, ends up echoing pathologizing narratives.[17]

Instead of being recognized as robust social institutions, race and class are treated as demographic variables attached to individual people and their immediate environments. T talk racializes and classes a range of behaviors and circumstances that go well beyond the ones we've looked at here. The narrative that T drives these domains contributes to the resurgent credibility of the idea that race is biological. Race and class return the favor for T, as familiar narratives of violence being endemic to poor or working class and black people, for example, go a long way toward covering holes and shortcomings in the scientific connections between T and aggression. The sociologist Avery F. Gordon describes race as a "ghostly matter," and its capacity to haunt as "a generalizable social phenomenon of great import. To study social life one must confront the ghostly aspects of it. This confrontation requires (or produces) a fundamental change in the way we know and make knowledge, in our mode of production." This entails reading around explicit statements to find what "was in the blind field, what was in the shadows, what only crazy people or powerless people saw." Similarly, the anthropologist Amade M'charek has characterized race as "an absent presence that oscillates between reality and nonreality because it is not a singular object but rather a pattern of various elements, some of which are made present and others absent." The "absent presence" formulation is especially apt for scientific work, where explicitly racialized narratives and racial stereotypes are typically avoided even as analyses borrow from and extend racial discourse and endorse white supremacist logics. The explicit absence of race enables racially pernicious material to travel across studies and into social life, while abdicating responsibility for the way that research justifies racial power structures. But the absence is illusory. Thus, in homage to Gordon, we have explored race as a "ghost variable" in research on T.[18]

The biologization of race is not new. Scholars have examined how new genetics models have ushered in a return to race as biology, but we have offered the first sustained attention to the role that hormones have played in this return. Endocrine research makes race biological in ways similar to but also different from the ways that genetic research does. One way that the genetics-race link is distinct is that genes are conceived as entities that are materially different among races, and those differences are seen as the root cause of race. Hormones, on the other hand, are chemically identical regardless of the body. Yet we have demonstrated that in some studies, especially some that explore population and/or racial differences in the neuroendocrine response to fatherhood, hormones are conceived as an evolutionary mechanism of racial differentiation, and as a signal that confirms theories of specifically racial trajectories of evolution.

The latest currents in feminist STS as well as social epidemiology conceptualize the intimate entangling of the biological and social through processes of "embodiment"; social class or race don't begin as biological entities, but they become biological as people literally ingest the material conditions of their lives. Recent studies on racial differences in T and behavior look entirely compatible with this approach, especially research on T and aggression. In his long history of research exploring high levels of T among young black men, Allan Mazur may seem to be enacting a concept of racial differences as the product of bodies that metabolize racism and structural violence. Given the current consensus that T rises in the face of a challenge, and that urban black communities are sites of great pressure and challenges, it seems reasonable to expect that you would find higher T in urban black communities. But race in Mazur's research is not a social structure with a history; it is a fixed property of the bodies and communities that generate the "environmental" challenges in the first place. Young black men are not under pressure from any broader social forces, the thinking goes, but rather from each other, and from the "fact" that they "dominate" black communities. In this version of race as embodiment, the social conditions that are internalized turn out to be the product of pathological race, rather than social institutions and history.[19]

A peculiar feature of race as it appears in that research is the way race is stabilized, and racial differences in T are likewise viewed as typological, within a model that's fundamentally about T's plasticity. Mazur's final crucial assumption is that any higher levels of T observed within a population of black men relative to a comparison population of white men will also correlate with (unmeasured) higher levels of aggression in black men. This rigid and relatively deterministic pattern remains consistently on one side of what Hammonds and Herzig describe as a toggling between the use of racial studies of hormones "to affirm absolute typological distinctions between bodies" and their use "to demonstrate the plasticity and continuity of an ambiguous spectrum."[20]

Biosocial frameworks have created space for a return to biologized views of social class, too. The idea that social class is a biologically grounded characteristic waxes and wanes, but it always lurks beneath the US ideal of self-making. There is a long tradition of belief that social position reflects the inherent "qualities" (meaning both characteristics and value) of individuals, though in the twenty-first century the understanding of socioeconomic class as biologically based is somewhat different than it was in earlier periods. Scholars have documented numerous ways that individual "biocitizens" have increasingly been seen as responsible for managing their own health, welfare, and reproduction as state responsibilities have shrunk over recent decades. These responsibilities are not so readily assumed and fulfilled by everyone: biological citizenship has been shown to be associated with sharp stratification of people within populations, especially those who are "othered" by virtue of racialization, disability, or extreme poverty. But work on biopolitics has generally not attended to how the broader structure of social class, not just resource deprivation experienced by those who are poor, may be woven into this devolution of state responsibilities, and into ideas about who is and isn't equipped to be an ideal biocitizen.[21]

Biological citizenship has arisen simultaneously with the "scientization" of the social sciences; incorporation of biological or biomedical theories and data has been an important strategy for elevating the status of these disciplines. At present, there is very little acknowledgment that a notion of social class as an expression of biological destiny is at work, especially in the US context. It is no surprise, then, that we lack any analysis of how this idea is crafted. At this historical moment, when the importance and volatility of class relations play such a powerful role in politics, it is especially urgent to shine some light on the casual contempt for poor and working-class people in the making of facts about T and behavior, whether they involve aggression, risk-taking, or reproduction. To be clear, not all of the work does this: anthropological studies of parenting among poor people often approach their reproductive decisions and family structures with respect and care. Our point here and elsewhere is not to generalize as if every tendency we describe in the literature on T can be found across all or even most studies, but to identify some important ways that T research operates as a field of power.

Hacking T with Social Theory

While people have been hacking gender with T, T can also be hacked with gender, or more precisely with gender theory. T talk, or discursive T, has already been deconstructed with feminist and gender theory in works that we've described and relied upon that show how beliefs and scientific facts about T and other hormones are integrated with discursive structures of gender and sexuality. As we've demonstrated throughout the book, lay beliefs and scientific facts about T are also woven through with discourses of race and class. Here, we want to focus on a different way in which social theory is crucial for understanding the chemical T, staying as close as possible to its materiality. Prior analyses have brought in social theory primarily to understand the intra-actions of scientific practices and the objects of study in enacting particular versions of hormones, T included. This continues to be urgent work, because the operations of gender systematically bias the scientific practices that engage T and other hormones. T is enacted via diverse medical and scientific engagements that materialize some versions of T while occluding or suppressing others. For example, the exact timing of collection for the blood, saliva, or muscle core samples from which T is measured is an intervention that both sifts material reality and affects it, as is the very act of collecting those samples. We've discussed experiments that involved fake money; relationships with "opponents" or "allies" who are only ever encountered through a single, staged computer game; programmable baby dolls that either do or don't respond to attempts at comforting; and more. These and other experimental situations are novel environments that don't merely reflect what T is doing "out there in the world" (as if there were some context-free version of T that could serve as the reference); they stage new relationships and intra-actions between T and a diverse cast of characters that include emotions, social relationships, and physical states.[22]

To continue the metaphor of research as a play, some habitual practices typecast T, setting up relations with a narrow and predictable set of characters that in turn limit the action that can unfold. Scientific knowledge is always a material-semiotic construction, but the attachment of T to gender ideology creates systematic biases. Most broadly, the habit of investigating T in men circumscribes the relations that can manifest scientifically. Even in studies that include women, broad associations of T with objects and psychological states culturally coded as masculine bring

T into proximity with aggression and competition, dominance, and money, for example.

All of this is broadly consistent with prior critical work on hormone research, though it extends and updates the examples. But we want to suggest a slight shift in attention, so that scientific practices become but one aspect of the social world with which the hormone engages.

Celia Roberts has observed that "Ernest Starling's choice of the classical Greek word *hormao* to form the root of the modern scientific term 'hormone' was felicitous. Etymologically, *hormao* means to excite or provoke. These are interesting actions: to provoke is to set something off, rather than to control or produce it." We share the interest that Roberts and others have shown in hormones' provocative capacities, and have underscored the resulting openness of T's relationships. But we have also turned the tables to consider that T is not just provocative but provoked by the social world. Reflecting on the verb "to provoke," Roberts remarks that its openness "leaves space for relations other than determinism. As provocative messengers, hormones create relations in articulation with other actors, retaining the potential to enact other relations in other times and spaces. Hormones are thus inherently (bio)political." Recognizing that hormones are not just agents provocateurs, but are also provoked by a wide variety of actors and actions that include social situations, boldly amplifies that conclusion.[23]

Thus while the fluidity and malleability of hormones have been topics of understanding since early endocrinology, the role of the social world in that fluidity and malleability has been undertheorized. If we begin by accepting that one of T's special material capacities is its responsiveness to social contexts and stimuli, then T also requires sophisticated understandings of the social. Understanding T as relational, and appreciating the specific material capacities of T to respond to social contexts, as well as to the meanings we assign to T and its somatic effects, thus shifts authority. Experts on social formations and forces have a clear role to play in theorizing how our bodies, via hormones, intra-act with social worlds. This insight extends to other (perhaps all) aspects of the body: bone formation, brain function, metabolism, the microbiome, and others are all areas where STS scholars (especially feminist scholars) have demonstrated intimate relations between the finest-grained aspects of body capacity and multiple levels of social dynamics and patterns. As Anne Fausto-Sterling has maintained, a fine-grained account of the materiality of the body, perhaps paradoxically, needs social theory. She continues, "While it is not reasonable, for example, to ask all biologists to become proficient in feminist theory or all feminist theorists to be proficient in cell biology, it is reasonable to ask each group of scholars to understand the limitations of knowledge obtained from working within a single discipline. Only non-hierarchical, multidisciplinary teams can devise more complete (or what Sandra Harding calls 'less false') knowledge." We agree wholeheartedly, and want to add a note on the current hierarchies of the disciplines that must come together in this way. Feminist and other social theorists currently rank far below biologists and other natural scientists, as well as experimental social scientists—especially those who engage biological variables like T. Those scientists have for decades recognized T's responsiveness to social situations, but as Sari van Anders has noted, they have typically relied on "commonsense," pre-theoretical ideas about social relations when they poke and prod T into response. The reason they fail to incorporate social theory is probably more a matter of power and the hierarchies of knowledge than it is of residual linearity in their modeling of T. Researchers who do empirical work with T are in general no more expert about gender and sexuality than lay people. Forty years of trenchant studies have demonstrated precisely how gendered and sexual ideologies and norms have been baked into our knowledge about "sex hormones," and yet, in all the hundreds of studies of T that we have read, we have never seen a single one of these STS analyses cited, except in van Anders's work.[24]

•   •   •

WHERE DOES THAT LEAVE T? For those who long for a tidier ending—a pronouncement that T "really" does these things over here, and does not do those things over there—by now it should be obvious that no easy answers about T will be forthcoming. If we have achieved our aims here, we have exposed readers to new insights into T and also to questions about it, showing it to be something far more interesting and complex than its traditional biographies have revealed. But we also hope to have shaken up not simply thinking about the hormone itself, but how ideas about T are used in the world to gloss entrenched social inequalities and to punt social concerns back to biology.

As we wrote this conclusion, an image formed of T as Atlas, bearing not a world but a worldview on his back. T has been made to bear a lot of weight. We know T talk will endure, as will the casual substitution in daily conversation of "testosterone" for men or masculinity and the "sex hormone" concept in research, but we hope that we have opened a space for new ways of thinking about T that might emerge alongside these, maybe taking up more room and gaining momentum as T's complexities are further elaborated. Instead of the titanic strength of Atlas, we hope we've suggested that T has other and better superpowers: a shape-shifting, moving, social molecule that serves as a dense transfer point for the micro-operations of biology and social relations of power at multiple levels.

Questions about biology and human nature are inextricable from moral and political debates about the value of human variations, the possibilities for equality, and the urgency and feasibility of social change. In some ways this book started with our prior work, but writing the book became urgent to us in the context of the regulation of women athletes' testosterone levels. It was in specific moments of that project that the human consequences of how we think about T came into sharp relief. In the cases heard by the Court of Arbitration for Sport, the IAAF advanced a supposedly authoritative but profoundly narrow and distorted "science of T" that rested on a static, binary, "sex hormone" view of T. That view in turn was used to justify exclusions and interventions against specific people: bodily, psychic, economic, and social harms that are both profound and impossible to fully assess. International sports bodies have so far tenaciously held on to their determination to bend T to their exclusionary regulations, but we hold out hope that better thinking about T might yet make a difference in that arena and so many others.

NOTES

Introduction: T Talk

1. "Testosterone," *This American Life*, August 30, 2002 (rebroadcast 2017), https://www.thisamericanlife.org/radio-archives/episode/220/testosterone.

2. [James Brown], "The Beast in Me," *GQ*, May 2002, 234.

3. Nelly Oudshoorn, *Beyond the Natural Body: An Archeology of Sex Hormones* (London: Routledge, 1994); Anne Fausto-Sterling, *Sexing the Body: Gender Politics and the Construction of Sexuality* (New York: Basic Books, 2000).

4. Oudshoorn, *Beyond*, 61.

5. Diana Long Hall, "Biology, Sex Hormones, and Sexism in the 1920's," *The Philosophical Forum* 5 (1973); Oudshoorn, *Beyond;* Fausto-Sterling, *Sexing;* Marianne van der Wijngaard, *Reinventing the Sexes: The Biomedical Construction of Femininity and Masculinity* (Bloomington: Indiana University Press, 1997); Chandak Sengoopta, "Glandular Politics: Experimental Biology, Clinical Medicine, and Homosexual Emancipation in Fin-de-Siècle Central Europe," *Isis* 89 (1998): 445–473.

6. Charles E. Brown-Séquard, "The Effects Produced on Man by Subcutaneous Injection of a Liquid Obtained from the Testicles of Animals," *Lancet* 137 (1889): 105–107.

7. "Is There an Elixir of Life?," *Boston Medical and Surgical Journal* 121 (1889): 167–168.

8. Victor Lespinasse, "Transplantation of the Testicle," *Chicago Medical Reader* 61 (1914): 1869–1870; Sengoopta, "Glandular"; Hall, "Biology," 81.

9. Ethan Blue, "The Strange Career of Leo Stanley: Remaking Manhood and Medicine at San Quentin State Penitentiary, 1913–1951," *Pacific Historical Review* 78 (2009): 210–241; Leo L. Stanley, "Testicular Substance Implantation," *Endocrinology* 5 (1921): 708–714; Serge Voronoff, *Life: A Study of the Means of*

Restoring Vital Energy and Prolonging Life (New York: Dutton, 1920); "Hopes to Find the Fountain of Youth in a Monkey Colony," *Lewiston Evening Journal,* July 18, 1923, 1.

10. Evelynn M. Hammonds and Rebecca M. Herzig, *The Nature of Difference: Sciences of Race in the United States from Jefferson to Genomics* (Cambridge, MA: MIT Press, 2009), 215. Most contemporary examinations of racial science either implicitly or explicitly assume that scientists have believed genes to be the foundational element through which supposedly biological race is built, but the documents presented by Herzig and Hammonds show that some scientists gave pride of place to hormones in this process. Both genes and hormones carried forward older ideas of "blood" as carrying the essence of racial or familial inheritance.

11. Oudshoorn, *Beyond;* Fausto-Sterling, *Sexing;* van den Wijngaard, *Reinventing;* Celia Roberts, *Messengers of Sex: Hormones, Biomedicine, and Feminism* (Cambridge, UK: Cambridge University Press, 2007); Ross Nehm and Rebecca Young, " 'Sex Hormones' in Secondary School Biology Textbooks," *Science and Education* 17 (2008): 1175–1190.

12. Nellie Bowles, "Push for Gender Equality in Tech? Some Men Say It's Gone Too Far," *New York Times,* September 24, 2017.

13. Andrew Sullivan, "The He Hormone," *New York Times Magazine,* April 2, 2000.

14. Jack van Honk, Geert-Jan Will, David Terburg, et al., "Effects of Testosterone Administration on Strategic Gambling in Poker Play," *Scientific Reports* 6 (2016), https://www.nature.com/articles/srep18096.pdf.

15. James McBride Dabbs and Mary Godwin Dabbs, *Heroes, Rogues, and Lovers: Testosterone and Behavior* (New York: McGraw-Hill, 2000); Robert A. Schug, "Understanding Disorders of Defiance, Aggression, and Violence: Oppositional Defiant Disorder, Conduct Disorder, and Antisocial Personality Disorder in Males," in *The Neuropsychology of Men: A Developmental Perspective,* ed. Charles M. Zaroff and Rik Carl D'Amato, 111–131 (New York: Springer, 2015).

16. John Hoberman, *Dopers in Uniform: The Hidden World of Police on Steroids* (Austin: University of Texas Press, 2017).

17. Quoted in Alexander Abad-Santos, "Gee Whiz, Saxby Chambliss Actually Said 'Hormones' Turn Troops into Rapists," *Atlantic,* June 4, 2013.

18. "Wilders: Migrant Men Are 'Islamic Testosterone Bombs,' " Al Jazeera, January 23, 2016, http://www.aljazeera.com/news/2016/01/wilders-migrant-men -islamic-testosterone-bombs-160123142600813.html.

19. Buck Gee and Denise Peck, *The Illusion of Asian Success: Scant Progress for Minorities in Cracking the Glass Ceiling from 2007–2015,* Ascend: Pan-Asian Leaders, n.d., https://c.ymcdn.com/sites/www.ascendleadership.org/resource /resmgr/research/TheIllusionofAsianSuccess.pdf.

20. Elizabeth A. Wilson, *Gut Feminism* (Durham, NC: Duke University Press, 2015).

21. Bruno Latour, "Why Has Critique Run out of Steam? From Matters of Fact to Matters of Concern," *Critical Inquiry* 30 (2004): 231.

22. Annemarie Mol, *The Body Multiple: Ontology in Medical Practice* (Durham, NC: Duke University Press, 2002).

23. Columbia University Institute for Social and Economic Research and Policy, "Special Initiative on Integrating Biology and Social Science Knowledge (BioSS)," April 26, 2019, http://iserp.columbia.edu/funding/special-initiative -integrating-biology-and-social-science-knowledge-bioss.

24. Some key works in feminist STS that denaturalize sex, gender, and sexuality include Anne Fausto-Sterling, *Myths of Gender: Biological Theories about Women and Men* (New York: Basic Books, 1985); Donna J. Haraway, *Primate Visions: Gender, Race, and Nature in the World of Modern Science* (New York: Routledge, 1989); Evelyn Fox Keller and Helen E. Longino, *Feminism and Science* (Oxford: Oxford University Press, 1996); Dorothy E. Roberts, *Killing the Black Body: Race, Reproduction, and the Meaning of Liberty* (New York: Pantheon Books, 1997); Sarah Blaffer Hrdy, *The Woman That Never Evolved* (Cambridge, MA: Harvard University Press, 1999); Sarah S. Richardson, *Sex Itself: The Search for Male and Female in the Human Genome* (Chicago: University of Chicago Press, 2013). For key work on "sex hormones," see Hall, "Biology"; Ruth Bleier, *Science and Gender: A Critique of Biology and Its Theories on Women* (New York: Pergamon Press, 1984); Oudshoorn, *Beyond;* Fausto-Sterling, *Sexing;* van den Wijngaard, *Reinventing;* Roberts, *Messengers;* Adele E. Clarke, *Disciplining Reproduction: Modernity, American Life Sciences, and "The Problems of Sex"* (Berkeley: University of California Press, 1998); Anne Fausto-Sterling, "The Bare Bones of Sex: Part 1—Sex and Gender," *Signs* 30, no. 2 (2005): 1491–1527; Wael Taha, Daisy Chin, Arnold I. Silverberg, et al., "Reduced Spinal Bone Mineral Density in Adolescents of an Ultra-Orthodox Jewish Community in Brooklyn," *Pediatrics* 107 (2001): E79; Jennifer R. Fishman, Laura Mamo, and Patrick R. Grzanka, "Sex, Gender, and Sexuality in Biomedicine," in *Handbook of Science and Technology Studies,* ed. Laurel Smith-Doerr, Clark Miller, Ulrike Felt, and Rayvon Fouche (Cambridge, MA: MIT Press, 2016), 379.

25. Cordelia Fine, *Testosterone Rex: Myths of Sex, Science, and Society* (New York: Norton, 2017). Fine focuses on T itself in two chapters. In a chapter on risk-taking, she masterfully deconstructs the notion that "risk" is a single domain of behavior, and that "risk-takers" will always eschew the safe path. In "The Hormonal Essence of the T-Rex," Fine surveys research across species that complicates the usual picture of a powerful male hormone that "serves to polarize the competitive behavior of the sexes."

26. For instance, STS scholar Anne Pollock has pointed out that while feminist critics have shown that environmental campaigns against endocrine disruptors mobilize fears of sexually non-normative bodies and behaviors, the same critics have nonetheless shared in the alarm about endocrine disruption. Maintaining that the alarm goes hand in hand with heteronormativity, Pollock encourages a vision of endocrine disruption as an opportunity for new forms of relation to emerge, eschewing an interest in "what is natural" in favor of "only what is, and what might be." Anne Pollock, "Queering Endocrine Disruption," in *Object-Oriented Feminism,* ed. Katherine Behar (Minneapolis: University of Minnesota

Press, 2016); Thomas Page McBee, *Amateur: A True Story about What Makes a Man* (New York: Scribner, 2018); Paul Preciado, *Testo Junkie: Sex, Drugs, and Biopolitics in the Pharmacopornographic Era* (New York: Feminist Press, 2013); Toby Beauchamp, "The Substance of Borders: Transgender Politics, Mobility, and U.S. State Regulation of Testosterone," *GLQ* 19 (2013): 57–78; Emilia Sanabria, *Plastic Bodies: Sex Hormones and Menstrual Suppression in Brazil* (Durham, NC: Duke University Press, 2016).

27. Evelynn M. Hammonds, "Straw Men and Their Followers: The Return of Biological Race," SSRC web forum on Race and Genomics, June 6, 2006, http://raceandgenomics.ssrc.org/Hammonds.

28. Hammonds and Herzig, *Nature*, 198.

29. Beth Loffreda and Claudia Rankine, *The Racial Imaginary: Writers on Race in the Life of the Mind* (Albany, NY: Fence Books, 2015), 19; for additional scholarship showing the persistent racialization of medico-scientific epistemologies, technologies, and institutions, see Troy Duster, *Backdoor to Eugenics* (New York: Routledge, 1991); Dorothy Roberts, *Fatal Invention: How Science, Politics, and Big Business Re-create Race in the Twenty-first Century* (New York: New Press, 2012); Ruha Benjamin, *People's Science: Bodies and Rights on the Stem Cell Frontier* (Stanford, CA: Stanford University Press, 2013); Alondra Nelson, *The Social Life of DNA: Race, Reparations, and Reconciliation After the Genome* (Boston: Beacon Press, 2016); Anthony Hatch, *Blood Sugar* (Minneapolis: University of Minnesota Press 2016).

30. Rebecca Jordan-Young, *Brain Storm: The Flaws in the Science of Sex Differences* (Cambridge, MA: Harvard University Press, 2010); Katrina Karkazis, *Fixing Sex: Intersex, Medical Authority, and Lived Experience* (Durham, NC: Duke University Press, 2008).

31. Katrina Karkazis, Rebecca M. Jordan-Young, Georgiann Davis, and Silvia Camporesi, "Out of Bounds? A Critique of the New Policies on Hyperandrogenism in Elite Female Athletes," *American Journal of Bioethics* 12 (2012): 3–16; Rebecca Jordan-Young and Katrina Karkazis, "Some of Their Parts: 'Gender Verification' and Elite Sports," *Anthropology News* (May 2012); Katrina Karkazis and Rebecca Jordan-Young, "The Harrison Bergeron Olympics," *American Journal of Bioethics* 13 (2013): 66–69; Rebecca Jordan-Young, Peter Sönksen, and Katrina Karkazis, "Sex, Health, and Athletes," *British Medical Journal* 348 (2014): g2926; Katrina Karkazis and Rebecca Jordan-Young, "Debating a Testosterone 'Sex Gap,'" *Science* 348 (2015): 858–860; Katrina Karkazis and Rebecca Jordan-Young, "The Powers of Testosterone: Obscuring Race and Regional Bias in the Regulation of Women Athletes," *Feminist Formations* 30 (2018): 1–39.

32. Joan W. Scott, "The Evidence of Experience," *Critical Inquiry* 17 (1991): 773–797.

33. Robert Proctor and Londa Schiebinger, eds., *Agnotology: The Making and Unmaking of Ignorance* (Stanford, CA: Stanford University Press, 2008), back cover; Nancy Tuana, "Coming to Understand: Orgasm and the Epistemology of Ignorance," *Hypatia* 19 (2004): 194–232.

1. Multiple Ts

1. US National Library of Medicine, "Testosterone," https://www.ncbi.nlm.nih
.gov/pubmedhealth/PMHT0027301, accessed August 3, 2018. Unfortunately, the
federal government discontinued PubMed Health, the NIH portal for systematic
reviews as well as consumer health information, on October 31, 2018. MedLine
Plus replaces this portal. T information is less expansive on this site, but just as
confusing and biased (e.g., "A testosterone test measures the amount of the male
hormone, testosterone, in the blood. Both men and women produce this hormone") (https://medlineplus.gov/ency/article/003707.htm).

2. Annemarie Mol, *The Body Multiple: Ontology in Medical Practice*
(Durham, NC: Duke University Press, 2003).

3. Benjamin Campbell and Michael Mbizo, "Reproductive Maturation,
Somatic Growth and Testosterone among Zimbabwe Boys," *Annals of Human
Biology* 33 (2006): 17–25; Liangpo Liu, Tongwei Xia, Xueqin Zhang, et al.,
"Biomonitoring of Infant Exposure to Phenolic Endocrine Disruptors Using Urine
Expressed from Disposable Gel Diapers," *Analytical and Bioanalytical Chemistry*
406, no. 20 (2014): 5049–5054.

4. James McBride Dabbs and Mary Godwin Dabbs, *Heroes, Rogues, and
Lovers: Testosterone and Behavior* (New York: McGraw-Hill, 2000), 6–7;
Thozhukat Sathyapalan, Ahmed Al-Qaissi, Eric S. Kilpatrick, et al., "Salivary
Testosterone Measurement in Women with and without Polycystic Ovary
Syndrome," *Scientific Reports* 7 (2017): 3589; Tom Fiers, Joris Delanghe, Guy
T'Sjoen, et al., "A Critical Evaluation of Salivary Testosterone as a Method for
the Assessment of Serum Testosterone," *Steroids* 86 (2014): 8.

5. Jerome Groopman, "Hormones for Men," *New Yorker*, July 22, 2002;
Virginia J. Vitzthum, Carol Worthman, Cynthia M. Beall, et al., "Seasonal and
Circadian Variation in Salivary Testosterone in Rural Bolivian Men," *American
Journal of Human Biology* 21 (2009): 762–768.

6. Emmanuele A. Jannini, Emiliano Screponi, Eleonora Carosa, et al., "Lack
of Sexual Activity from Erectile Dysfunction Is Associated with a Reversible
Reduction in Serum Testosterone," *International Journal of Andrology* 22 (1999):
385–392; Shawn N. Geniole, Brian M. Bird, Erika L. Ruddick, and Justin M.
Carré, "Effects of Competition Outcome on Testosterone Concentrations in
Humans: An Updated Meta-Analysis," *Hormones and Behavior* 92 (2017):
37–50; Shawn N. Geniole, Justin M. Carre, and Cheryl M. McCormick, "State,
Not Trait, Neuroendocrine Function Predicts Costly Reactive Aggression in Men
after Social Exclusion and Inclusion," *Biological Psychology* 87 (2011): 137–145;
C. Martyn Beaven, Will G. Hopkins, Kier T. Hansen, et al., "Dose Effect of
Caffeine on Testosterone and Cortisol Responses to Resistance Exercise,"
International Journal of Sport Nutrition and Exercise Metabolism 18 (2008):
131–141; Gary G. Gordon, Kurt Altman, A. Louis Southren, Emanuel Rubin, and
Charles S. Lieber, "Effect of Alcohol (Ethanol) Administration on Sex-Hormone
Metabolism in Normal Men," *New England Journal of Medicine* 295 (1976):
793–797; Frederick C. W. Wu, Abdelouahid Tajar, Stephen R. Pye, et al.,

"Hypothalamic-Pituitary-Testicular Axis Disruptions in Older Men Are Differentially Linked to Age and Modifiable Risk Factors: The European Male Aging Study," *Journal of Clinical Endocrinology and Metabolism* 93 (2008): 2737–2745; Sari M. van Anders, Richard M. Tolman, and Brenda L. Volling, "Baby Cries and Nurturance Affect Testosterone in Men," *Hormones and Behavior* 61 (2012): 31–36; Kimberly A. Cote, Cheryl M. McCormick, Shawn N. Geniole, Ryan P. Renn, and Stacey D. MacAulay, "Sleep Deprivation Lowers Reactive Aggression and Testosterone in Men," *Biological Psychology* 92 (2013): 249–256.

7. Mazen Shihan, Ahmed Bulldan, and Georgios Scheiner-Bobis, "Non-Classical Testosterone Signaling Is Mediated by a G-Protein-Coupled Receptor Interacting with Gnα11," *Biochimica et Biophysica Acta* 1843 (2014): 1172–1181.

8. "Testosterone, Total, Available, and Free," Mayo Clinic Laboratories, 2011, https://www.mayomedicallaboratories.com/test-catalog/2011/Clinical+and+Interpretive/83686.

9. Joëlle Taieb, Bruno Mathian, Françoise Millot, et al., "Testosterone Measured by 10 Immunoassays and by Isotope-Dilution Gas Chromatography-Mass Spectrometry in Sera from 116 Men, Women, and Children," *Clinical Chemistry* 49 (2003): 1381–1395; David A. Herold and Robert L. Fitzgerald, "Immunoassays for Testosterone in Women: Better than a Guess?," *Clinical Chemistry* 49 (2003): 1250–1251. We initially thought the title of this last piece, an editorial, was tongue-in-cheek, but the writers of the editorial argued that with clinical information, a physician's educated guess would in fact be more accurate than available immunoassays.

10. Douglas A. Granger, Elizabeth A. Shirtcliff, Alan Booth, Katie T. Kivlighan, and Eve B. Schwartz, "The 'Trouble' with Salivary Testosterone," *Psychoneuroendocrinology* 29 (2004): 1229–1240.

11. Groopman, "Hormones."

2. Ovulation

1. Peter Casson, M. S. Lindsay, Margareta D. Pisarska, Sandra A. Carson, and John E. Buster, "Dehydroepiandrosterone Supplementation Augments Ovarian Stimulation in Poor Responders: A Case Series," *Human Reproduction* 15 (2000): 2129–2132. For a review of what was known at that time regarding acupuncture for treatment of female infertility, including ovulation induction, see Raymond Chang, Pak H. Chung, and Zev Rosenwaks, "Role of Acupuncture in the Treatment of Female Infertility," *Fertility and Sterility* 78 (2002): 1149–1153.

2. For a sample of the brief descriptions of early stages of egg development in medical texts, see Jonathan S. Berek, ed., *Berek and Novak's Gynecology*, 12th ed. (Philadelphia: Wolters Kluwer Health/Lippincott Williams & Wilkins, 2012); Barbara L. Hoffman, John O. Schorge, Karen D. Bradshaw, et al., eds., *Williams Gynecology*, 3rd ed. (New York: McGraw-Hill Education, 2016); F. Gary Cunningham, Kenneth J. Leveno, Steven L. Bloom, et al., eds., *Williams Obstetrics*, 23rd ed. (New York: McGraw-Hill Professional, 2009).

3. Hoffman et al., eds., *Williams Gynecology.*

4. Nelly Oudshoorn, *Beyond the Natural Body: An Archeology of Sex Hormones* (London: Routledge, 1994); Anne Fausto-Sterling, *Sexing the Body: Gender Politics and the Construction of Sexuality* (New York: Basic Books, 2000); Marianne van der Wijngaard, *Reinventing the Sexes: The Biomedical Construction of Femininity and Masculinity* (Bloomington: Indiana University Press, 1997).

5. Whalen as quoted in Wijngaard, *Reinventing,* 42–43; Ross Nehm and Rebecca Young, "'Sex Hormones' in Secondary School Biology Textbooks," *Science and Education* 17 (2008): 1175–1190; Fausto-Sterling, *Sexing.*

6. Bruce S. McEwen, Ivan Lieberburg, Claude Chaptal, and Lewis C. Krey, "Aromatization: Important for Sexual Differentiation of the Neonatal Rat Brain," *Hormones and Behavior* 9 (1977): 249–263; Daniela C. C. Gerardin and Oduvaldo C. M. Pereira, "Reproductive Changes in Male Rats Treated Perinatally with an Aromatase Inhibitor," *Pharmacology, Biochemistry, and Behavior* 71 (2002): 301–305; Lewis C. Krey, Ivan Lieberburg, Neil MacLusky, and Bruce S. McEwen, "Aromatization and Development of Responsiveness of the Brain to Gonadal Steroids," in *Development of Responsiveness to Steroid Hormones: Advances in the Biosciences,* ed. Alvin M. Kaye and Myra Kaye (Oxford, UK: Pergamon, 1980), 423–431.

7. Casson et al., "Dehydroepiandrosterone Supplementation."

8. Kirsty A. Walters, "Role of Androgens in Normal and Pathological Ovarian Function," *Reproduction* 149 (2015): R197; Norbert Gleicher, Andrea Weghofer, and David H. Barad, "The Role of Androgens in Follicle Maturation and Ovulation Induction: Friend or Foe of Infertility Treatment?," *Reproductive Biology and Endocrinology* 9 (2011): 116. This discussion of testosterone or other androgens and ovulation illustrates why and how we move between the language of "women" and "females." Some of the data and all of the clinical situations involve humans only, so we use the term "women"; but when we're discussing general theories and data that are meant to apply across species, we use the term "females."

9. Walters, "Role of Androgens."

10. Amir Wiser, Ofer Gonen, Yehudith Ghetler, et al., "Addition of Dehydroepiandrosterone (DHEA) for Poor-Responder Patients before and during IVF Treatment Improves the Pregnancy Rate: A Randomized Prospective Study," *Human Reproduction* 25 (2010): 2496–2500.

11. Pasquale Patrizio, Alberto Vaiarelli, Paolo E. Levi Setti, et al., "How to Define, Diagnose and Treat Poor Responders? Responses from a Worldwide Survey of IVF Clinics," *Reproductive Biomedicine Online* 30 (2015): 581–592. For women who have the necessary enzyme to convert DHEA to T, treatment with DHEA is preferable to directly treating with T for a few possible reasons. First, it's hard to calibrate the precise amount of T that's necessary, and raising levels too high can tip women into territory where they experience side effects, including direct negative effects on developing follicles. Second, because DHEA is a precursor to many other steroids, treating with DHEA affects a wider array of hormone levels and may be less disruptive to beneficial relative proportions of

hormones involved in fertility, including androgens, estrogens, progestogens, LH, and FSH.

12. Kayhan Yakin and Bulent Urman, "DHEA as a Miracle Drug in the Treatment of Poor Responders; Hype or Hope?," *Human Reproduction* 26 (2011): 1941–1944; Helen E. Nagels, Josephine R. Rishworth, Charalampos S. Siristatidis, and Ben Kroon, "Androgens (Dehydroepiandrosterone or Testosterone) for Women Undergoing Assisted Reproduction," *Cochrane Database of Systematic Reviews* 11 (2015): CD009749.

13. Annemarie Mol, *The Body Multiple: Ontology in Medical Practice* (Durham, NC: Duke University Press, 2002).

14. Robert Proctor and Londa Schiebinger, eds., *Agnotology: The Making and Unmaking of Ignorance* (Stanford, CA: Stanford University Press, 2008); Emily Martin, "The Egg and the Sperm: How Science Has Constructed a Romance Based on Stereotypical Male-Female Roles," *Signs* 16 (1991): 485–501; Sarah Richardson, *Sex Itself : The Search for Male and Female in the Human Genome* (Chicago: University of Chicago Press, 2013).

3. Violence

1. William Greider, "Army Recounts Testimony of Calley Unit," *Washington Post,* March 16, 1971; Stuart Auerbach, "Army Studies Tests of Aggressiveness," *Washington Post,* March 16, 1971; Leo E. Kreuz and Robert M. Rose, "Assessment of Aggressive Behavior and Plasma Testosterone in a Young Criminal Population," *Psychosomatic Medicine* 34 (1972): 321–332.

2. Evelynn M. Hammonds and Rebecca M. Herzig, *The Nature of Difference: Sciences of Race in the United States from Jefferson to Genomics* (Cambridge, MA: MIT Press, 2009); Amade M'charek, "Beyond Fact or Fiction: On the Materiality of Race in Practice," *Cultural Anthropology* 28 (2013): 423.

3. Christopher Mims, "Strange but True: Testosterone Alone Does Not Cause Violence," *Scientific American,* July 5, 2007.

4. Allan Mazur, "Testosterone Is High among Young Black Men with Little Education," *Frontiers in Sociology* 1 (2016): 1; Charles Ramsey, "Steroid Use Has Been on the Rise in Philadelphia," *Subject to Debate: A Newsletter of the Police Executive Research Forum* 26 (2012); John Hoberman, *Dopers in Uniform: The Hidden World of Police on Steroids* (Austin: University of Texas Press, 2017).

5. John Archer, "Testosterone and Human Aggression: An Evaluation of the Challenge Hypothesis," *Neuroscience and Biobehavioral Reviews* 30 (2006): 320; James Dabbs, Gregory J. Jurkovic, and Robert L. Frady, "Salivary Testosterone and Cortisol among Late Adolescent Male Offenders," *Journal of Abnormal Child Psychology* 19 (1991): 469–478; James Dabbs and Mary Dabbs, *Heroes, Rogues, and Lovers: Testosterone and Behavior* (New York: McGraw-Hill, 2000).

6. Examples of placebo-controlled studies that find no link include Ray Tricker, Richard Casaburi, Thomas W. Storer, et al., "The Effects of Supraphysiological Doses of Testosterone on Angry Behavior in Healthy Eugonadal Men—A Clinical Research Center Study," *Journal of Clinical Endocrinology and Metabo-*

lism 81 (1996): 3754–3758; Daryl B. O'Connor, John Archer, and Frederick W. C. Wu, "Effects of Testosterone on Mood, Aggression, and Sexual Behavior in Young Men: A Double-Blind, Placebo-Controlled, Cross-Over Study," *Journal of Clinical Endocrinology and Metabolism* 89 (2004): 2837–2845. There are a few exceptions to the general pattern that exogenous T in placebo-controlled trials does not increase aggression, but those findings should be viewed with some skepticism. For instance, Pope, Kouri, and Hudson report that exogenous T increased aggression, but the aggression measure that was linked to higher T in their study was a computer game (the point-subtraction aggression paradigm, PSAP), which was only linked to T in a subsample of twenty-seven men, while scores on aggression measures that were used in the entire sample of fifty-six, including a validated measure of aggressive actions and feelings and an assessment by significant others, were not associated with T. Harrison G. Pope, Elena M. Kouri, and James I. Hudson, "Effects of Supraphysiologic Doses of Testosterone on Mood and Aggression in Normal Men—A Randomized Controlled Trial," *Archives of General Psychiatry* 57 (2000): 133–140. Another study found that exogenous T was linked to more "aggressive" play on the PSAP, but only among men who scored high on a "dominance" measure or low on a "self-control" measure. Even so, the effects were quite modest: "Collectively, the trait dominance by drug condition and trait self-control by drug condition interactions accounted for 8.8% of the variance in aggressive behavior." Justin M. Carré, Shawn N. Geniole, Triana L. Ortiz, et al., "Exogenous Testosterone Rapidly Increases Aggressive Behavior in Dominant and Impulsive Men," *Biological Psychiatry* 82 (2017): 249–256.

7. One thread we don't take up in this chapter is the notion that prenatal hormone exposures "hardwire" the brain for more greater aggressive behavior. Jordan-Young examined this claim at length, and found it rested on a fundamentally flawed research model combined with selective omission of negative findings in key studies. Rebecca Jordan-Young, *Brain Storm: The Flaws in the Science of Sex Differences* (Cambridge, MA: Harvard University Press, 2010), esp. 210–213, 227–228.

8. "Stress, Aggression, and Male Hormones," *New York Times*, May 7, 1972.

9. Ronald L. Goldfarb and Linda R. Singer, "Maryland's Defective Delinquency Law and the Patuxent Institution," *Bulletin of the Menninger Clinic* 34 (1970): 223–235.

10. Kreuz and Rose, "Assessment." When these studies on prisoners, T, and aggressive criminality began, there was not yet a scientific consensus that there are special ethical concerns involved when conducting research with prisoners. In 1976, new ethical guidelines laid out principles that would govern such research, but it's unlikely that these principles had much impact on the research done in this particular vein, because there was a strong emphasis in the guidelines on studies that might pose physical risks. By the time the studies by James Dabbs and colleagues (described later in this chapter) were under way, researchers would have had to certify compliance with these rules. None of the literature we've examined on T in prisoners addresses ethical issues, but we think it's worth reflecting on the way that prisoners have been and still are a vulnerable research population, in

part because of the unique pressures to participate that might actually affect treatment by guards and others, or might be perceived to do so.

11. Kreuz and Rose, "Assessment," 327.

12. Laura Bradley, "In the New X-Files, 'I Want to Believe' Has Lost Its Meaning," *Slate,* January 26, 2016.

13. Kreuz and Rose, "Assessment," 327.

14. Auerbach, "Army"; Rose quoted in Auerbach, "Army."

15. Brenda Remmes, "James McBride Dabbs, Jr., Son of James and Edith Dabbs," *Everything Happens at the Crossroads,* 2012, https://dabbscrossroads .blogspot.com/2012/09/james-mcbride-dabbs-jr-son-of-james-and.html; Dabbs and Dabbs, *Heroes,* back jacket.

16. James Dabbs, Robert Frady, Timothy Carr, and Norma Besch, "Saliva Testosterone and Criminal Violence in Young-Adult Prison-Inmates," *Psychosomatic Medicine* 49 (1987): 174–182.

17. Jean-Claude Dreher, Simon Dunne, Agnieszka Pazderska, et al., "Testosterone Causes Both Prosocial and Antisocial Status-Enhancing Behaviors in Human Males," *Proceedings of the National Academy of Sciences* 113 (2016): 11633; John T. Whitehead and Steven P. Lab, *Juvenile Justice: An Introduction* (New York: Routledge, 2015).

18. Dabbs, Jurkovic, and Frady, "Salivary," 470; John Archer, "The Influence of Testosterone on Human Aggression," *British Journal of Psychology* 82 (1991): 1–28.

19. Daniel Goleman, "Aggression in Men: Hormone Levels Are a Key," *New York Times,* July 17, 1990.

20. Dabbs quoted in Goleman, "Aggression."

21. Goleman, "Aggression."

22. James Dabbs and Robin Morris, "Testosterone, Social Class, and Antisocial Behavior in a Sample of 4,462 Men," *Psychological Science* 1 (1990): 209–211.

23. Alair MacLean, "The Stratification of Military Service and Combat Exposure, 1934–1994," *Social Science Research* 40 (2011): 336–348.

24. Dabbs and Dabbs, *Heroes,* 150.

25. M'charek, "Beyond."

26. Allan Mazur and Theodore A. Lamb, "Testosterone, Status, and Mood in Human Males," *Hormones and Behavior* 14 (1980): 236–246; Nancy Krieger and George D. Smith, "'Bodies Count,' and Body Counts: Social Epidemiology and Embodying Inequality," *Epidemiologic Reviews* 26 (2004): 92.

27. John C. Wingfield, Robert E. Hegner, Alfred M. Dufty, and Gregory F. Ball, "The 'Challenge Hypothesis': Theoretical Implications for Patterns of Testosterone Secretion, Mating Systems, and Breeding Strategies," *American Naturalist* 136 (1990): 833.

28. Allan Mazur and Alan Booth, "Testosterone and Dominance in Men," *Behavioral and Brain Sciences* 21 (1998): 353–363. Mazur and Booth didn't mention Wingfield's research directly, but several of the commentaries published alongside their article did. The original hypothesis was developed to explain patterns in species with mating seasons, which is sometimes seen as creating

special problems for applying this model to humans. Others, however, have suggested that this is not a fundamental problem, but simply would lead to different specific predictions. For instance, in a response to Mazur and Booth, Rui Oliveira writes: "As the human species is considered to be monogamous and does not present a breeding seasonality, the challenge hypothesis would predict human male T levels to respond sharply to social challenges. In fact, the data presented by M&B provide further evidence for the challenge hypothesis; T rises in response to a competitive match, as if in anticipation of the challenge." Rui Oliveira, "Of Fish and Men: A Comparative Approach to Androgens and Social Dominance—Commentary / Mazur & Booth: Testosterone and Dominance," *Behavioral and Brain Sciences* 21 (1998): 383. While it is conventional to distinguish between Mazur and Booth's "biosocial model" and Wingfield's "challenge hypothesis," there were connections from the beginning. Two commentaries on Mazur and Booth's review noted that their model seemed consistent with the challenge hypothesis, and Mazur and Booth concurred, calling the challenge hypothesis "an excellent opportunity for students of animal and human behavior to integrate them into a general model that applies across vertebrate species"; Mazur and Booth, "Testosterone," 387.

29. Archer, "Testosterone," 320; Mazur and Booth, "Testosterone."

30. Richard E. Nisbett, "Violence and U.S. Regional Culture," *American Psychologist* 48 (1993): 441–449; Richard E. Nisbett and Dov Cohen, *Culture of Honor* (Boulder, CO: Westview Press, 1996); Dov Cohen, Richard E. Nisbett, Brian F. Bowdle, and Norbert Schwarz, "Insult, Aggression, and the Southern Culture of Honor: An 'Experimental Ethnography,'" *Journal of Personality and Social Psychology* 70 (1996): 945–960; Mazur and Booth, "Testosterone," 360.

31. Elijah Anderson as quoted in Mazur and Booth, "Testosterone," 360.

32. Lee Ellis and Helmuth Nyborg, "Racial / Ethnic Variations in Male Testosterone Levels," *Steroids* 57 (1992): 72–75.

33. Mazur and Booth, "Testosterone."

34. Mazur and Booth, "Testosterone," 360 (emphasis in original).

35. Mazur and Booth, "Testosterone," 354.

36. Allan Mazur, "Biosocial Models of Deviant Behavior among Army Veterans," *Biological Psychology* 41 (1995): 291, 282, 289.

37. Consider, for example, Carré and colleagues' long-term evaluation of an early intervention program for "aggressive-disruptive" kindergartners. The original sample was mostly white. But when researchers sought to explore the program's effects on adult T dynamics and "aggressive" reactions to provocation, they only followed up with the black subsample. Likewise, Scerbo and Kolko's exploration of testosterone and cortisol in "aggressive-disruptive children" focuses on a sample that is 75 percent African American. Conversely, a Web of Science search for recent studies (2010 to mid-2018) with the keywords "aggression," "testosterone," and "human" yielded mostly reports of laboratory paradigms of dominance and status, all of which employ exclusively or majority white samples. Justin M. Carré, Anne-Marie R. Iselin, Keith M. Welker, Ahmad R. Hariri, and Kenneth A. Dodge, "Testosterone Reactivity to Provocation Mediates the Effect of Early Intervention on Aggressive Behavior," *Psychological Science* 25

(2014): 1140–1146; Angela Scarpa Scerbo and David J. Kolko, "Salivary Testosterone and Cortisol in Disruptive Children—Relationship." *Journal of the American Academy of Child and Adolescent Psychiatry* 33 (1994): 1174–1184.

38. Mazur, "Testosterone"; Allan Mazur, *Biosociology of Dominance and Deference* (New York: Rowman and Littlefield, 2005).

39. Compare Pranjal H. Mehta and Robert A. Josephs, "Testosterone and Cortisol Jointly Regulate Dominance: Evidence for a Dual-Hormone Hypothesis," *Hormones and Behavior* 58 (2010): 898–906; Allan Mazur and Alan Booth, "Testosterone Is Related to Deviance in Male Army Veterans, but Relationships Are Not Moderated by Cortisol," *Biological Psychology* 96 (2014): 72–76; Jack van Honk, Eddie Harmon-Jones, Barak E. Morgan, and Dennis J. L. G. Schutter, "Socially Explosive Minds: The Triple Imbalance Hypothesis of Reactive Aggression," *Journal of Personality* 78 (2010): 67–94.

40. Van Honk et al., "Socially," 69.

41. Steven E. Barkan and Michael Rocque, "Socioeconomic Status and Racism as Fundamental Causes of Street Criminality," *Critical Criminology* 26 (2018): 211; Anthony Walsh and Ilhong Yun, "Examining the Race, Poverty, and Crime Nexus Adding Asian Americans and Biosocial Processes," *Journal of Criminal Justice* 59 (2018): 42. For a particularly insightful analysis of the differential racialization of Asian Americans and African Americans, see Claire Jean Kim, "The Racial Triangulation of Asian Americans." *Politics and Society* 27 (1999): 105–138.

42. Amade M'charek, "Tentacular Faces and Generous Methods for Studying Race," Public Lecture, New School for Social Research, October 18, 2017. https://events.newschool.edu/event/anthropology_lecture_-_amade_mcharek# .W0ZIyX4nat9.

43. Michelle Alexander, *The New Jim Crow: Mass Incarceration in the Age of Colorblindness* (New York: New Press, 2010); Tess Borden, "Every 25 Seconds: The Human Toll of Criminalizing Drug Use in the United States," Human Rights Watch and American Civil Liberties Union, 2016, https://www.hrw.org/report /2016/10/12/every-25-seconds/human-toll-criminalizing-drug-use-united-states; "BOP Statistics: Inmate Offenses," accessed February 3, 2019, https://www.bop .gov/about/statistics/statistics_inmate_offenses.jsp.

44. Banu Subramaniam, *Ghost Stories for Darwin: The Science of Variation and the Politics of Diversity* (Champagne, IL: University of Illinois Press, 2014).

4. Power

1. Amy Cuddy, "Your Body Language May Shape Who You Are," TEDGlobal, 2012, https://www.ted.com/talks/amy_cuddy_your_body_language_shapes_who _you_are; Dana Carney, Amy Cuddy, and Andy Yap, "Power Posing: Brief Nonverbal Displays Affect Neuroendocrine Levels and Risk Tolerance," *Psychological Science* 21 (2010): 1363–1368.

2. Amy Cuddy, *Presence: Bringing Your Boldest Self to Your Biggest Challenges* (New York: Little, Brown, 2015).

3. Carney, Cuddy, and Yap, "Power," 1363; John Archer, "Testosterone and Human Aggression: An Evaluation of the Challenge Hypothesis," *Neuroscience and Biobehavioral Reviews* 30 (2006): 319–345; Allan Mazur and Alan Booth, "Testosterone and Dominance in Men," *Behavioral and Brain Sciences* 21 (1998): 353–363; Carney, Cuddy, and Yap, "Power," 1363–1364.

4. Robert Sapolsky, *The Trouble with Testosterone: And Other Essays on the Biology of the Human Predicament* (New York: Scribner, 1998).

5. Carney, Cuddy, and Yap, "Power," 1363.

6. Carney, Cuddy, and Yap, "Power," 1364.

7. Carney, Cuddy, and Yap, "Power," 1367.

8. Steven J. Stanton, "The Essential Implications of Gender in Human Behavioral Endocrinology Studies," *Frontiers in Behavioral Neuroscience 5* (2011): 1–3.

9. Eva Ranehill, Anna Dreber, Magnus Johannesson, et al., "Assessing the Robustness of Power Posing No Effect on Hormones and Risk Tolerance in a Large Sample of Men and Women," *Psychological Science* 26 (2015): 653–656.

10. Dana Carney, Amy Cuddy, and Andy Yap, "Review and Summary of Research on the Embodied Effects of Expansive (vs. Contractive) Nonverbal Displays," *Psychological Science* 26 (2015): 657–663.

11. Joe Simmons and Uri Simonsohn, "Power Posing: Reassessing the Evidence Behind the Most Popular TED Talk," Data Colada, May 8, 2015, http://datacolada.org/37; Andrew Gelman and Kaiser Fung, "The Power of the 'Power Pose,'" *Slate,* January 19, 2016.

12. Dana Carney, "My Position on Power Poses," 2016, http://faculty.haas .berkeley.edu/dana_carney/pdf_My%20position%20on%20power%20poses.pdf.

13. Cuddy as quoted in Jesse Singal and Melissa Dahl, "Here Is Amy Cuddy's Response to Critiques of Her Power-Posing Research," *New York Magazine,* September 30, 2016; Carney, Cuddy, and Yap, "Power," 1363.

14. Susan Dominus, "When the Revolution Came for Amy Cuddy," *New York Times Magazine,* October 22, 2017.

15. Kristopher M. Smith and Coren L. Apicella, "Winners, Losers, and Posers: The Effect of Power Poses on Testosterone and Risk-Taking Following Competition," *Hormones and Behavior* 92 (2017): 172–181.

16. Smith and Apicella, "Winners," 172.

17. Amy J. Cuddy, "Feeling Powerless Is Not Being Powerless," Momicon 2016, https://www.youtube.com/watch?v=-1i1Bcuhib.

18. Eva Ranehill, Anna Dreber, Magnus Johannesson, et al., as posted on Data Colada, May 8, 2015, http://datacolada.org/37.

19. Stanton, "Essential."

20. Sheryl Sandberg, *Lean In: Women, Work, and the Will to Lead* (New York: Knopf, 2013), 34; Aja Romano, "Michelle Obama on Sheryl Sandberg's Lean In Philosophy: "That Shit Doesn't Work All the Time!," *Vox,* December 3, 2018.

21. Evelynn M. Hammonds and Rebecca M. Herzig, *The Nature of Difference: Sciences of Race in the United States from Jefferson to Genomics* (Cambridge, MA: MIT Press, 2009), 199.

22. Victoria Pitts-Taylor, "Plastic Brain: Neoliberalism and the Neuronal Self," *Health: Interdisciplinary Studies in Health, Illness and Medicine* 14, no. 6 (2010): 635; Victoria Pitts-Taylor, *The Brain's Body: Neuroscience and Corporeal Politics* (Durham, NC: Duke University Press, 2016); Sigrid Schmitz, "Sex/Gender in the Cerebral Subject: Feminist Reflections on Modern Neuro-Cultures," keynote lecture, First International Conference of Neurogenderings: Critical Studies of the Sexed Brain, Uppsala, Sweden, March 25, 2010; Nikolas Rose, "The Death of the Social? Re-figuring the Territory of Government," *Economy and Society* 25 (2016): 327–356.

23. Hammonds and Herzig, *Nature,* 198; Emilia Sanabria, *Plastic Bodies: Sex Hormones and Menstrual Suppression in Brazil* (Durham, NC: Duke University Press, 2016), 114.

24. Nelly Oudshoorn, *Beyond the Natural Body: An Archeology of Sex Hormones* (New York: Routledge, 1994); Anne Fausto-Sterling, *Sexing the Body: Gender Politics and the Construction of Sexuality* (New York: Basic Books, 2000); Sari van Anders, "Beyond Masculinity: Testosterone, Gender/Sex, and Human Social Behavior in a Comparative Context," *Frontiers in Neuroendocrinology* 34 (2013): 198–210; Cordelia Fine, *Testosterone Rex: Myths of Sex, Science, and Society* (New York: Norton, 2017); Toby Beauchamp, "The Substance of Borders: Transgender Politics, Mobility, and US State Regulation of Testosterone," *GLQ* 19, no. 1 (2012): 57–78.

25. Paul B. Preciado, *Testo Junkie: Sex, Drugs, and Biopolitics in the Pharmacopornographic Era* (New York: Feminist Press, 2013), 395.

26. Patricia Hill Collins, *Black Feminist Thought: Knowledge, Consciousness, and the Politics of Empowerment* (Boston, MA: Unwin Hyman, 1990).

27. Joan C. Williams, Katherine W. Phillips, and Erika V. Hall, *Double Jeopardy? Gender Bias Against Women of Color in Science* (Berkeley: UC Hastings Law School, 2014), http://www.uchastings.edu/news/articles/2015/01/double-jeopardy-report.pdf.

28. Beverly Smith, online comment on "The Power of Presence," *On Point,* WBUR, February 3, 2016.

29. Ryan Grim, "The Transcript of Sandra Bland's Arrest Is as Revealing as the Video," *Huffington Post,* July 22, 2015.

30. Grim, "Transcript"; David Montgomery, "Sandra Bland Was Threatened with Taser, Police Video Shows," *New York Times,* July 21, 2015; St. John Barned-Smith and Leah Binkovitz, "Trooper Who Pulled Over Bland Placed on Administrative Duty," *Houston Chronicle,* July 17, 2015.

5. Risk-Taking

1. Marvin Kusmierz, "Anna Edson Taylor (1839–1921)," *Saginaw Bay-Journal,* http://bay-journal.com/bay/1he/people/fp-taylor-annie.html; Eric Grundhauser, "Annie Edson Taylor's 1901 Retirement Plan: Go over Niagara Falls in a Barrel," *Atlas Obscura,* https://www.atlasobscura.com/articles/annie-edson-taylors-1901-retirement-plan-go-over-niagara-falls-in-a-barrel; Dwight

Whalen, *The Lady Who Conquered Niagara: The Annie Edson Taylor Story* (Brewer, ME: EGA Books, 1990); "Woman Goes over Niagara in a Barrel," *New York Times*, October 25, 1901.

2. "Woman," 1.

3. Annie Edson Taylor, *Over the Falls: Annie Edson Taylor's Story of Her Trip: How the Horseshoe Fall Was Conquered*, 1902, https://archive.org/stream /overfallsannieed00tayluoft/overfallsannieed00tayluoft_djvu.txt.

4. Cordelia Fine, *Testosterone Rex: Myths of Sex, Science, and Society* (New York: Norton, 2017), 15.

5. Coren Apicella, Anna Dreber, Benjamin Campbell, et al., "Testosterone and Financial Risk Preferences," *Evolution and Human Behavior* 29 (2008): 384–390; Coren Apicella, Anna Dreber, and Johanna Mollerstrom, "Salivary Testosterone Change Following Monetary Wins and Losses Predicts Future Financial Risk-Taking," *Psychoneuroendocrinology* 39 (2014): 58–64; Paola Sapienza, Luigi Zingales, and Dario Maestripieri, "Gender Differences in Financial Risk Aversion and Career Choices Are Affected by Testosterone," *Proceedings of the National Academy of Sciences* 106 (2009): 15268–15273; Pablo Brañas-Garza and Aldo Rustichini, "Organizing Effects of Testosterone and Economic Behavior: Not Just Risk Taking," *PLOS ONE*, 2011, https://doi .org/10.1371/journal.pone.0029842; Pranjal H. Mehta, Keith M. Welker, Samuele Zilioli, and Justin M. Carré, "Testosterone and Cortisol Jointly Modulate Risk Taking," *Psychoneuroendocrinology* 56 (2015): 88–99; Eric Stenstrom, Gad Saad, Marcelo V. Nepomuceno, and Zack Mendenhall, "Testosterone and Domain-Specific Risk: Digit Ratios (2D:4D and Rel2) as Predictors of Recreational, Financial, and Social Risk-Taking Behaviors," *Personality and Individual Differences* 51 (2011): 412–416; Roderick E. White, Stewart Thornhill, and Elizabeth Hampson, "Entrepreneurs and Evolutionary Biology: The Relationship Between Testosterone and New Venture Creation," *Organizational Behavior and Human Decision Processes* 100 (2006): 21–34; Ellen Garbarino, Robert Slonim, and Justin Sydnor, "Digit Ratios (2D:4D) as Predictors of Risky Decision Making for Both Sexes," *Journal of Risk and Uncertainty* 42 (2011): 1–26; Henrik Cronqvist, Alessandro Previtero, Stephan Siegel, and Roderick E. White, "The Fetal Origins Hypothesis in Finance: Prenatal Environment, the Gender Gap, and Investor Behavior," *Review of Financial Studies* 29 (2016): 739–786; Shinichi Kamiya, Y. Han (Andy) Kim, and Soohyun Park, "The Face of Risk: CEO Facial Masculinity and Firm Risk," forthcoming in *European Financial Management*, https://ssrn.com/abstract=2557038, http://dx.doi.org/10.2139/ssrn .2557038.

6. Minda Zetlin, "5 Things the Smartest Leaders Know About Risk-Taking," Inc.com, https://www.inc.com/minda-zetlin/5-things-the-smartest-leaders-know -about-risk-taking.html; "risk-taking," Merriam Webster, https://www.merriam -webster.com/dictionary/risk-taking.

7. Apicella et al., "Testosterone," 387.

8. White, Thornhill, and Hampson, "Entrepreneurs," 21, 23, 25.

9. White, Thornhill, and Hampson, "Entrepreneurs," 23; James Dabbs, Denise de la Rue, and Pam Williams, "Testosterone and Occupational Choice—Actors,

Ministers, and Other Men," *Journal of Personality and Social Psychology* 59 (1990): 1262–1263.

10. White, Thornhill, and Hampson, "Entrepreneurs," 23.

11. Sandra E. Black, Paul J. Devereux, Petter Lundborg, and Kaveh Majlesi, "On the Origins of Risk-Taking," National Bureau of Economic Research working paper 21332, July 2015, 14, 18–19.

12. Antoine Bechara, Hanna Damasio, Daniel Tranel, and Antonio R. Damasio, "The Iowa Gambling Task and the Somatic Marker Hypothesis: Some Questions and Answers," *Trends in Cognitive Sciences* 9 (2005): 159–162.

13. Jack van Honk, Dennis J. L. G. Schutter, Erno J. Hermans, et al., "Testosterone Shifts the Balance between Sensitivity for Punishment and Reward in Healthy Young Women," *Psychoneuroendocrinology* 29 (2004): 937–943; Anna E. Goudriaan, Bruno Lapauw, Johannes Ruige, et al., "The Influence of High-Normal Testosterone Levels on Risk-Taking in Healthy Males in a 1-Week Letrozole Administration Study," *Psychoneuroendocrinology* 35 (2010): 1416–1421; Steven J. Stanton, Scott H. Liening, and Oliver C. Schultheiss, "Testosterone Is Positively Associated with Risk Taking in the Iowa Gambling Task," *Hormones and Behavior* 59 (2011): 252–256; William H. Overman and Allison Pierce, "Iowa Gambling Task with Non-Clinical Participants: Effects of Using Real Virtual Cards and Additional Trials," *Frontiers in Psychology* 4 (2013): 935. Tom Hildebrandt, James W. Langenbucher, Adrianne Flores, Seth Harty, and Heather A. Berlin, "The Influence of Age of Onset and Acute Anabolic Steroid Exposure on Cognitive Performance, Impulsivity, and Aggression in Men," *Psychology of Addictive Behaviors* 28 (2014): 1096–1104; Kelly L. Evans and Elizabeth Hampson, "Does Risk-Taking Mediate the Relationship between Testosterone and Decision-Making on the Iowa Gambling Task?," *Personality and Individual Differences* 61–62 (2014): 57–62.

14. Van Honk et al., "Testosterone," 939.

15. Jack van Honk, Geert-Jan Will, David Terburg, et al., "Effects of Testosterone Administration on Strategic Gambling in Poker Play," *Scientific Reports* 6 (2016): 1.

16. Van Honk et al., "Testosterone"; Goudriaan et al., "Influence"; Stanton et al., "Testosterone"; Overman and Pierce, "Iowa"; Hildebrandt et al., "Influence"; Evans and Hampson, "Risk-Taking."

17. Stanton et al., "Testosterone."

18. John Coates and Joe Herbert, "Endogenous Steroids and Financial Risk Taking on a London Trading Floor," *Proceedings of the National Academy of Sciences* 105 (2008): 6167–6172.

19. Katrin Bennhold, "Where Would We Be If Women Ran Wall Street?," *New York Times,* February 1, 2009.

20. Coates and Herbert, "Endogenous," 6167. They describe their hypotheses and key findings this way: "Specifically, we predicted that testosterone would rise on days when traders made an above-average gain in the markets, and cortisol would rise on days when traders were stressed by an above-average loss. Our data confirmed the first prediction but suggested that cortisol responds more to uncertainty of return than to loss."

21. John Coates, *The Hour between Dog and Wolf: How Risk Taking Transforms Us, Body and Mind* (New York: Penguin, 2012), 181.

22. Coates and Herbert, "Endogenous."

23. Though they share similar weaknesses, the fate of Coates and Herbert's study differs significantly from that of the power posing study by Carney and colleagues that we discussed in Chapter 4. Where that study, and Cuddy's promotion of it, have been widely skewered, this study remains virtually unscathed. Both Cuddy and Coates have taken their research to an enormous public stage, with bestselling books and high-profile media appearances. We can't say why Coates and Herbert's work managed to get past the reviewers and editor at the prestigious *PNAS*, let alone all the smart bloggers out there who so eagerly picked apart Cuddy's work.

24. John Coates, "The Biology of Risk," *New York Times,* June 7, 2014; Olivia Solon, "Testosterone Is to Blame for Financial Market Crashes, Says Neuroscientist," *Wired,* July 13, 2012.

25. John M. Coates, Mark Gurnell, and Zoltan Sarnyai, "From Molecule to Market: Steroid Hormones and Financial Risk-Taking," *Philosophical Transactions of the Royal Society B: Biological Sciences* 365 (2010): 331–343.

26. Coates and Herbert, "Endogenous," 6167.

27. Jens O. Zinn, "The Meaning of Risk-Taking—Key Concepts and Dimensions," *Journal of Risk Research,* https://doi.org/10.1080/13669877.2017 .1351465, 3; Yuping Jia, Laurence van Lent, and Yachang Zeng, "Masculinity, Testosterone, and Financial Misreporting," *Journal of Accounting Research* 52 (2014): 1195–1246; Michael P. Haselhuhn and Elaine M. Wong, "Bad to the Bone: Facial Structure Predicts Unethical Behaviour," *Proceedings of the Royal Society of Biological Sciences* 279 (2012): 571–576.

28. Fine, *Testosterone,* 117; Sari van Anders, Katherine L. Goldey, Terri Conley, Daniel J. Snipes, and Divya A. Patel, "Safer Sex as the Bolder Choice: Testosterone Is Positively Correlated with Safer Sex Behaviorally Relevant Attitudes in Young Men," *Journal of Sexual Medicine* 9 (2012): 727–734; John C. Rosenblitt, Hosanna Soler, Stacy E. Johnson, and David M. Quadagno, "Sensation Seeking and Hormones in Men and Women: Exploring the Link," *Hormones and Behavior* 40 (2001): 396–402; Eric Stenstrom, Gad Saad, Marcelo V. Nepomuceno, and Zack Mendenhall, "Testosterone and Domain-Specific Risk: Digit Ratios (2D:4D and rel2) as Predictors of Recreational, Financial, and Social Risk-Taking Behaviors," *Personality and Individual Differences* 51 (2011): 412–416.

29. Marvin Zuckerman, *Behavioral Expressions and Biosocial Bases of Sensation Seeking* (New York: Cambridge University Press, 1994). The scale is used in these studies: Rosenblitt et al., "Sensation"; Bernhard Fink, Aicha Hamdaoui, Frederike Wenig, and Nick Neave, "Hand-Grip Strength and Sensation Seeking," *Personality and Individual Differences* 49 (2010): 789–793; Benjamin Campbell, Anna Dreber, Coren Apicella et al., "Testosterone Exposure, Dopaminergic Reward and Sensation-Seeking in Young Men," *Physiology and Behavior* 99 (2010): 451–456; Martin Voracek, Ulrich S. Tran, and Stefan G. Dressler, "Digit Ratio (2D:4D) and Sensation Seeking: New Data and Meta-Analysis," *Personality and Individual Differences* 48 (2010): 72–77.

Instead of looking at individual components of risk, we could step back and understand risk-taking as a form of decision-making. While studies on risk-taking and T tend to take risk behavior or "risk propensities" as a coherent package, neurobiologist Antonio Damasio seeks to understand the interplay of emotion, bodily sensations, and higher level cognitive processes in risk-taking or avoidance. Important convergences between Damasio's work and studies of T and risk-taking suggest it is fruitful to look at how he thinks about risk. Both, for instance, focus less on the logical risk calculations people make and more on risk as an affective or emotional orientation. As Damasio writes, "Emotion and feeling, along with the covert physiological machinery underlying them, assist us with the daunting task of predicting an uncertain future and planning our actions accordingly." In the T and risk literature, language such as "appetite for risk," "risk propensity," and the "drive," "tolerance," or "willingness" for risk point in the same direction. Antonio Damasio, *Descartes' Error: Emotion, Reason, and the Human Brain* (New York: Grosset/Putnam, 1994), xiii.

30. Fine, *Testosterone;* Stenstrom et al., "Testosterone," 413.

31. Zinn, "Meaning," 3, 2.

32. Karen Messing and Jeanne Mager Stellman, "Sex, Gender and Women's Occupational Health: The Importance of Considering Mechanism," *Environmental Research Volume* 101 (2006): 149–162.

33. Fine, *Testosterone;* CDC, "Pregnancy Mortality Surveillance System," November 9, 2017, https://www.cdc.gov/reproductivehealth/maternalinfanthealth /pmss.html.

34. Susan R. Fisk, Brennan J. Miller, and Jon Overton, "Why Social Status Matters for Understanding the Interrelationships between Testosterone, Economic Risk-Taking, and Gender," *Sociology Compass* 11 (2017): e12452, 1.

35. Fisk et al., "Why," 8.

36. Thekla Morgenroth, Cordelia Fine, Michelle K. Ryan, and Anna E. Genat, "Sex, Drugs, and Reckless Driving: Are Measures Biased toward Identifying Risk-Taking in Men?," *Social Psychological and Personality Science*, 2017, https://doi.org/10.1177/1948550617722833.

37. Fisk et al., "Why."

38. Sari M. van Anders, "Chewing Gum Has Large Effects on Salivary Testosterone, Estradiol, and Secretory Immunoglobulin A Assays in Women and Men," Psychoneuroendocrinology 35 (2010): 305–309.

39. See, for example, Matthew Pearson and Burkhard C. Schipper, "Menstrual Cycle and Competitive Bidding," July 5, 2012, http://dx.doi.org/10.2139/ssrn .1441665; Brañas-Garza and Rustichini, "Organizing"; Voracek et al., "Digit"; Sapienza et al., "Gender"; Daphna Joel and Ricardo Tarrasch, "The Risk of a Wrong Conclusion: On Testosterone and Gender Differences in Risk Aversion and Career Choices," *Proceedings of the National Academy of Sciences* 107 (2010): E19.

40. Examples of studies that use the dual hormone hypothesis are Mehta et al., "Testosterone" and van Honk et al., "Testosterone." Susman et al. posit that T moderates effect of "cortisol reactivity," and the "coupling" hypothesis is described in Shirtcliff et al.: Elizabeth J. Susman, Melissa K. Peckins, Jacey L.

Bowes, and Lorah D. Dorn, "Longitudinal Synergies between Cortisol Reactivity and Diurnal Testosterone and Antisocial Behavior in Young Adolescents," *Development and Psychopathology* 29 (2017): 1353–1369; Elizabeth A. Shirtcliff, Andrew R. Dismukes, Kristine P. Marceau, et al., "A Dual-Axis Approach to Understanding Neuroendocrine Development," *Developmental Psychobiology* 57 (2015): 643–653.

6. Parenting

1. Lee T. Gettler, Thomas McDade, Alan Feranil, and Christopher Kuzawa, "Longitudinal Evidence That Fatherhood Decreases Testosterone in Human Males," *Proceedings of the National Academy of Sciences* 108 (2011): 16194–16199. The first study to demonstrate lower T among human fathers compared with non-fathers was Anne E. Storey, Carolyn J. Walsh, Roma L. Quinton, and Katherine E. Wynne-Edwards, "Hormonal Correlates of Paternal Responsiveness in New and Expectant Fathers," *Evolution and Human Behavior* 21 (2000): 79–95. For additional examples of cross-sectional research, see Peter B. Gray, Sonya M. Kahlenberg, Emily S. Barrett, Susan F. Lipson, and Peter T. Ellison, "Marriage and Fatherhood Are Associated with Lower Testosterone in Males," *Evolution and Human Behavior* 23 (2002): 193–220; Peter B. Gray, Chi-Fu Jeffrey Yang, and Harrison G. Pope Jr., "Fathers Have Lower Salivary Testosterone Levels than Unmarried Men and Married Non-Fathers in Beijing, China," *Proceedings of the Royal Society of London B: Biological Sciences* 273 (2006): 333–339; Martin Muller, Frank Marlowe, Revocatus Bugumba, and Peter Ellison, "Testosterone and Paternal Care in East African Foragers and Pastoralists," *Proceedings of the Royal Society of London B: Biological Sciences* 276 (2009): 347–354; Christopher W. Kuzawa, Lee T. Gettler, Martin N. Muller, Thomas W. McDade, and Alan B. Feranil, "Fatherhood, Pairbonding and Testosterone in the Philippines," *Hormones and Behavior* 56 (2009): 429–435. Studies aren't always clear about whether they only included men who were biological fathers, and that distinction may or may not matter, depending on how researchers conceptualize the underlying dynamics, as will become clear later in this chapter.

2. The quote is from Bhaskar Prasad, "Decrease in Testosterone Level after Fatherhood May Protect Men from Chronic Diseases," *International Business Times News,* September 13, 2011; Ian Sample, "Being a Dad Makes Less of a Man," *Guardian,* September 13, 2011; Mike Swain, "Why Dads' Sex Drive is Stuck in Reverse; Testosterone Levels Plummet 34%," *Daily Mirror,* September 13, 2011; Kate Clancy, "Parenting is Not Just for the Ladies: On Testosterone, Fatherhood, and Why Lower Hormones Are Good for You," *Scientific American,* September 16, 2011.

3. Ellison quoted in Pam Belluck, "Fatherhood Cuts Testosterone, Study Finds, for Good of the Family," *New York Times,* September 12, 2011.

4. Alex Williams, "Testosterone Study Has Fathers Questioning Their Manhood," *New York Times,* September 16, 2011.

5. Kuzawa et al., "Fatherhood, Pairbonding."

6. Williams, "Testosterone."

7. Gettler quoted in Bonnie Rochman, "Dads Have Less Testosterone," *Time,* December 7, 2011; Kuzawa quoted in Sample, "Being"; Gettler quoted in Jennifer Welsh, "Fatherhood Lowers Testosterone, Keeps Dads at Home," *Scientific American,* September 12, 2011; Lee T. Gettler, "Direct Male Care and Hominin Evolution: Why Male-Child Interaction Is More than a Nice Social Idea," *American Anthropologist* 112 (2010): 7–21; Lee T. Gettler, Chris Kuzawa, Thomas McDade, and Alan Feranil, "Fatherhood, Childcare, and Testosterone: Study Authors Discuss the Details," *Scientific American,* October 5, 2011.

8. Gettler, "Direct."

9. Sarah Blaffer Hrdy, "Care and Exploitation of Nonhuman Primate Infants by Conspecifics Other than the Mother," in *Advances in the Study of Behavior,* ed. Jay S. Rosenblatt, Robert A. Hinde, Evelyn Shaw, and Colin Beer (New York: Academic, 1976): 6:101–158.

10. Gettler, "Direct," 8. It's worth noting that the literature on parenting and T could be analyzed for a pervasive "adaptationist bias," meaning that re-searchers and theorists tend to assume that any evolved traits are adaptations, when adaptation is only one mechanism of natural selection. For a lucid explana-tion of adaptationist bias, see Elisabeth Anne Lloyd, *The Case of the Female Orgasm: Bias in the Science of Evolution* (Cambridge, MA: Harvard University Press, 2005). In this vein, and related to our discussion later in this chapter of how studies of T in fathers incorrectly assert links between higher T and various traits like libido and aggression, the cultural anthropologist Daniel Segal pointed out in a blog post that the press coverage and some authors' comments from the study by Gettler and colleagues rests on an "adaptationist fable" that lower T will lead to monogamy among new fathers. He doesn't take on the claim that lower T also makes men less likely to hurt their children, perhaps because Gettler and colleagues mostly emphasize the reduced "mating" investment that might be related to lower T. Daniel A. Segal, "Testosterone and Culture: A Comment on Another Adaptationist Fable," Shake Well Before Using, September 19, 2011, http://daniel-segal.blogspot.com/2011/09/headline-on-september-12-read.html.

11. Gettler, "Direct," 9; C. Owen Lovejoy, "The Origin of Man," *Science* 211 (1981): 341–350.

12. Frank W. Marlowe, "Hunting and Gathering: The Human Sexual Division of Foraging Labor," *Cross-Cultural Research* 41 (2007): 170–195. The works described in this section generally use the term "sex" to refer to the way that foraging niches and behaviors were or were not specific to males and females in early human history. While we would typically use the word "gender" to describe male and female behaviors, we have opted to follow the lead of the evolutionary scholars here, in part for consistency, and in part because the early development of gender is precisely what is at stake in this discussion.

13. David Epstein, *The Sport Gene: Inside the Science of Extraordinary Athletic Performance* (New York: Portfolio, 2014), 73.

14. Robert Trivers, "Parental Investment and Sexual Selection," in *Sexual Selection and the Descent of Man* (Chicago: Aldine, 1972), 55; Kim Hill and Hillard Kaplan, "Life History Traits in Humans: Theory and Empirical Studies,"

Annual Review of Anthropology 28 (1999): 402. For an in-depth analysis of how evolutionary theory has incorporated the assumptions and categories of capitalist economic theory, see Richard Lewontin and Richard Levins, *Biology as Ideology: The Doctrine of DNA* (New York: Harper Perennial, 1991).

15. Sarah K. C. Holtfrerich, Katharina A. Schwarz, Christian Sprenger, Luise Reimers, and Esther K. Diekhof, "Endogenous Testosterone and Exogenous Oxytocin Modulate Attentional Processing of Infant Faces," *PLOS ONE* 11 (2016): e0166617.

16. Gettler et al., "Fatherhood, Childcare," n.p.

17. John Archer, "Testosterone and Human Aggression: An Evaluation of the Challenge Hypothesis," *Neuroscience and Biobehavioral Reviews* 30 (2006): 320; James Dabbs, Gregory J. Jurkovic, and Robert L. Frady, "Salivary Testosterone and Cortisol among Late Adolescent Male Offenders," *Journal of Abnormal Child Psychology* 19 (1991): 469–478; James Dabbs and Mary Dabbs, *Heroes, Rogues, and Lovers: Testosterone and Behavior* (New York: McGraw-Hill, 2000); James M. Dabbs, "Testosterone Measurements in Social and Clinical-Psychology," *Journal of Social and Clinical Psychology* 11 (1992): 309.

18. Abdulmaged M. Traish and Andre T. Guay, "Are Androgens Critical for Penile Erections in Humans? Examining the Clinical and Preclinical Evidence," *Journal of Sexual Medicine* 3 (2006): 382–404; Sari van Anders, "Beyond Masculinity: Testosterone, Gender/Sex, and Human Social Behavior in a Comparative Context," *Frontiers in Neuroendocrinology* 34 (2013): 203; Shalendar Bhasin, Paul Enzlin, Andrea Coviello, and Rosemary Basson, "Sexual Dysfunction in Men and Women with Endocrine Disorders," *Lancet* 369 (2007): 597–611. But regarding the threshold concept, see Andrea M. Isidori, Elisa Giannetta, Daniele Gianfrilli, et al., "Effects of Testosterone on Sexual Function in Men: Results of a Meta-Analysis," *Clinical Endocrinology* 63 (2005): 381–394.

19. Kaye Wellings, Martine Collumbien, Emma Slaymaker, et al., "Sexual and Reproductive Health 2: Sexual Behaviour in Context: A Global Perspective," *The Lancet* 368 (2006): 1706–1728; Peter B. Gray and Kermyt G. Anderson, *Fatherhood: Evolution and Human Paternal Behavior* (Cambridge, MA: Harvard University Press, 2012).

20. The "problem" of "matrifocal" households was a staple of social science literature in the mid- to late-twentieth century, in which "matrifocal" signaled an implicitly pathological black family structure. Almost two decades before the challenge hypothesis was applied to humans, anthropologists Patricia Draper and Henry Harpending coined the terms "dad strategy" and "cad strategy," with the former being the "invested" father who opts for high quality offspring, and the latter being the one who leaves the children behind in "matrifocal" households. They explained: "Male children born into matrifocal households exhibit at adolescence a complex of aggression, competition, low male parental investment, and derogation of females and femininity, while females show early expression of sexual interest and assumption of sexual activity, negative attitudes toward males, and poor ability to establish long-term relationships with one male." Among children reared with fathers, though, they claim that boys are more interested in "manipulation of nonhuman aspects of the environment" than in dominance and

competition, and girls are slower to develop sexually and are more interested in finding a male provider. Resonance with the challenge hypothesis is strong: unstable environments, aggression, competition, and low parental investment cluster. The only thing missing is the T. Like sociologists' culture of poverty theories, Draper and Harpending's summation reflects broad cultural prejudice about deviance versus adherence to the norms of heterosexual, middle-class, white households. Patricia Draper and Henry Harpending, "Father Absence and Reproductive Strategy: An Evolutionary Perspective," *Journal of Anthropological Research* 38 (1982): 255.

21. Ramya M. Rajagopalan, Alondra Nelson, and Joan H. Fujimura, "Race and Science in the Twenty-First Century," in *The Handbook of Science and Technology Studies,* 4th ed., ed. Ulrike Felt, Rayvon Fouché, Clark A. Miller, and Laurel Smith-Doerr (Cambridge, MA: MIT Press, 2016), 353.

22. Muller et al., "Testosterone and Paternal," 348.

23. Muller et al., "Testosterone and Paternal," 352.

24. Peter Gray, "Human Fatherhood Is Diverse: They Do What They Need to Do," Fatherhood, September 16, 2016, https://fatherhood.global/human -fatherhood-diverse; Jo Jones and William D. Mosher, "Fathers' Involvement with Their Children: United States, 2006–2010," *National Health Statistics Reports* 71 (2013): 1–21. The "Fathers' Involvement" report compares non-Hispanic white men, non-Hispanic black men, and Hispanic men of any "race."

25. Gray and Anderson, *Fatherhood;* Gray, "Human Fatherhood"; Peter B. Gray, "Failing Our Fathers," *Psychology Today,* April 6, 2015, https://www .psychologytoday.com/us/blog/the-evolving-father/201504/failing-our-fathers.

26. Jennifer S. Mascaro, Patrick D. Hackett, and James K. Rilling, "Testicular Volume Is Inversely Correlated with Nurturing-Related Brain Activity in Human Fathers," *Proceedings of the National Academy of Sciences* 110 (2013): 15746; Belluck, "Fatherhood"; Sarah Zhang, "Better Fathers Have Smaller Testicles," *Nature,* September 9, 2013; Brian Alexander, "Aw Nuts! Nurturing Dads Have Smaller Testicles, Study Shows," NBC News, September 9, 2013. Researchers intent on countering the normativity of this body of work have their work cut out for them. One place to start would be to connect their work with data that complicate ideas about men who "choose" to be "absent" and "uninvolved" fathers. As Charles Blow has said, the high rate of US black children who don't live with their fathers "represents more than [men's] choice. It exists in a social context, one at odds with the corrosive mythology about black fathers." Charles M. Blow, "Black Dads Are Doing Best of All," *New York Times,* December 21, 2017.

27. Robert H. MacArthur and Edward O. Wilson, *The Theory of Island Biogeography* (Princeton, NJ: Princeton University Press, 1967).

28. J. Philippe Rushton and Anthony F. Bogaert, "Race Differences in Sexual-Behavior: Testing an Evolutionary Hypothesis," *Journal of Research in Personality* 21 (1987): 546. Rushton and Bogaert use "Orientals." Celia Roberts, *Messengers of Sex: Hormones, Biomedicine, and Feminism* (New York: Cambridge University

Press, 2007); Claire Jean Kim, "The Racial Triangulation of Asian Americans," *Politics & Society* 27 (1999): 106.

29. Lee Ellis, "Criminal Behavior and *r/K* Selection: An Extension of Gene-Based Evolutionary Theory," *Personality and Individual Differences* 9 (1988): 701; Lee Ellis, "Sex Hormones, *R/K* Selection, and Victimful Criminality," *Mankind Quarterly* 29 (1989): 329–340.

30. Lee Ellis and Helmuth Nyborg, "Racial/Ethnic Variations in Male Testosterone Levels: A Probable Contributor to Group Differences in Health," *Steroids* 57 (1992): 72–75; Helmuth Nyborg, *Hormones, Sex, and Society* (Westport, CT: Praeger, 1994); J. Philippe Rushton, "Race, Genetics, and Human Reproductive Strategies," *Genetic, Social and General Psychology Monographs* 122 (1996): 21–53. Evidence that Ellis and Nyborg were specifically looking to support the idea that human races are differentially selected on the "*r/K* continuum" is circumstantial but compelling. For one thing, Ellis had already published three papers promoting Rushton's theory before he teamed up with Nyborg to analyze the army data. Second, he and Nyborg thank Rushton in the acknowledgments of their paper. Third, they couch their analysis of racial differences in T in a discussion of health disparities, specifically in prostate cancer rates; two years earlier, another promoter of Rushton's theory, the British psychologist Richard Lynn, had offered racial differences in prostate cancer as "indirect evidence" of racial difference in *r/K* selection. Richard Lynn, "Testosterone and Gonadotrophin Levels and *r/K* Reproductive Strategies," *Psychological Reports* 67 (1990): 1203–1206.

31. Allan Mazur and Alan Booth, "Testosterone and Dominance in Men," *Behavioral and Brain Sciences* 21 (1998): 353–363.

32. Helmuth Nyborg, "Migratory Selection for Inversely Related Covariant T-, and IQ-Nexus Traits: Testing the IQ/T-Geo-Climatic-Origin Theory by the General Trait Covariance Model," *Personality and Individual Differences* 55 (2013): 272; Ellis and Nyborg, "Racial/Ethnic," 74; John M. Hoberman, *Darwin's Athletes: How Sport Has Damaged Black America and Preserved the Myth of Race* (Boston: Houghton Mifflin, 1997).

33. Sari M. van Anders, Katherine L. Goldey, and Patty X. Kuo, "The Steroid/Peptide Theory of Social Bonds: Integrating Testosterone and Peptide Responses for Classifying Social Behavioral Contexts," *Psychoneuroendocrinology* 36 (2011): 1266.

34. Kathy Trang, "Mapping Out a Feminist Bioscience: Interview with Sari van Anders (Part 1)," Foundation for Psychocultural Research, May 1, 2015, https://thefpr.org/mapping-out-a-feminist-bioscience-interview-with-sari-van -anders-part-1.

35. Van Anders, "Beyond," 198.

36. Van Anders et al., "Steroid," 1272.

37. Lee T. Gettler, "Becoming DADS: Considering the Role of Cultural Context and Developmental Plasticity for Paternal Socioendocrinology," *Current Anthropology* 57 (2016): S40.

38. Gettler, "Becoming," S46.

7. Athleticism

1. Natasha Singer, "Does Testosterone Build a Better Athlete?," *New York Times*, August 10, 2006. Illustrating Granger's point that T alone isn't enough to make someone athletic, multiple studies show T levels among high-level male athletes, including elite weightlifters and top-flight track and field athletes, are not significantly different from those of non-athletes, including: Joan Carles Arce, Mary Jane De Souza, Linda S. Pescatello, and Anthony A. Luciano, "Subclinical Alterations in Hormone and Semen Profile in Athletes," *Fertility and Sterility* 59 (1993): 398–404; Philippe Passelergue, Annie R. Robert, and G. Lac, "Salivary Cortisol and Testosterone Variations during an Official and a Simulated Weight-lifting Competition," *International Journal of Sports Medicine* 16 (1995): 298–303; Carmelo Bosco, Roberto Colli, Roberto Bonomi, Serge P. Von Duvillard, and Atko Viru, "Monitoring Strength Training: Neuromuscular and Hormonal Profile," *Medicine and Science in Sports and Exercise* 32 (2000): 202–208; Juha P. Ahtiainen, Arto Pakarinen, Markku Alen, William J. Kraemer, and Keijo Häkkinen, "Muscle Hypertrophy, Hormonal Adaptations during Strength Training in Strength-Trained and Strength Development and Untrained Men," *European Journal of Applied Physiology* 89 (2003): 555–563; Mikel Izquierdo, Javier Ibañéz, Keijo Häkkinen, et al., "Maximal Strength and Power, Muscle Mass, Endurance and Serum Hormones in Weightlifters and Road Cyclists," *Journal of Sports Sciences* 22 (2004): 465–478.

2. On muscle mass, strength, and endurance, see Shalender Bhasin, Thomas Storer, Nancy Berman, et al., "The Effects of Supraphysiologic Doses of Testosterone on Muscle Size and Strength in Normal Men," *New England Journal of Medicine* 335 (1996): 1–7; Christina Wang, Ronald S. Swerdloff, Ali Iranmanesh, et al., "Transdermal Testosterone Gel Improves Sexual Function, Mood, Muscle Strength, and Body Composition Parameters in Hypogonadal Men," *Journal of Clinical Endocrinology and Metabolism* 85 (2000): 2839–2853; Thomas W. Storer, Lynne Magliano, Linda Woodhouse, et al., "Testosterone Dose-Dependently Increases Maximal Voluntary Strength and Leg Power, but Does Not Affect Fatigability or Specific Tension," *Journal of Clinical Endocrinology and Metabolism* 88 (2003): 1478–1485.

3. Examples of studies showing higher T correlating with performance include Marco Cardinale and Michael H. Stone, "Is Testosterone Influencing Explosive Performance?," *Journal of Strength and Conditioning Research* 20 (2006): 103–107; William J. Kraemer, John F. Patton, Scott E. Gordon, et al., "Compatibility of High-Intensity Strength and Endurance Training on Hormonal and Skeletal-Muscle Adaptations," *Journal of Applied Physiology* 78 (1995): 976–989; Bosco et al., "Monitoring Strength." Studies showing no relationship between T and performance include Passelergue, Robert, and Lac, "Salivary"; Blair T. Crewther, Liam P. Kilduff, Christian J. Cook, et al., "The Acute Potentiating Effects of Back Squats on Athlete Performance," *Journal of Strength and Conditioning Research* 25 (2011): 3319–3325; Cláudio Balthazar, Marcia Garcia, and Regina Spadari-Bratfisch, "Salivary Concentrations of Cortisol and Testosterone and Prediction of Performance in a Professional Triathlon Competition," *Stress* 15 (2012): 495–502;

Blair T. Crewther, Liam Kilduff, Christian Cook, et al., "Relationships between Salivary Free Testosterone and the Expression of Force and Power in Elite Athletes," *Journal of Sports Medicine and Physical Fitness* 52 (2012): 221–227. For studies that find a negative relationship between T and performance, see: Blair T. Crewther, Zbigniew Obminski, and Christian Cook, "The Effect of Steroid Hormones on the Physical Performance of Boys and Girls during an Olympic Weightlifting Competition," *Pediatric Exercise Science* 28 (2016): 580–587; Brandon K. Doan, Robert U. Newton, William J. Kraemer, Young-Hoo Kwon, and Timothy P. Scheet, "Salivary Cortisol, Testosterone, and T/C Ratio Responses during a 36-Hole Golf Competition," *International Journal of Sports Medicine* 28 (2007): 470–479; Izquierdo et al., "Maximal."

4. Christopher M. Gaviglio, Blair T. Crewther, Liam P. Kilduff, Keith A. Stokes, and Christian J. Cook, "Relationship between Pregame Concentrations of Free Testosterone and Outcome in Rugby Union," *International Journal of Sports Physiology and Performance* 9 (2014): 324–331; Blair T. Crewther, Tim Lowe, Robert P. Weatherby, Nicholas Gill, and Justin Keogh, "Neuromuscular Performance of Elite Rugby Union Players and Relationships with Salivary Hormones," *Journal of Strength and Conditioning Research* 23 (2009): 2046–2053; Blair T. Crewther, Christian Cook, Chris Gaviglio, Liam Kilduff, and Scott Drawer, "Baseline Strength Can Influence the Ability of Salivary Free Testosterone to Predict Squat and Sprinting Performance," *Journal of Strength and Conditioning Research* 26 (2012): 261–268; William J. Kraemer, Duncan N. French, Nigel J. Paxton, et al., "Changes in Exercise Performance and Hormonal Concentrations over a Big Ten Soccer Season in Starters and Nonstarters," *Journal of Strength and Conditioning Research* 18 (2004)."

5. Karen Choong, Kishore M. Lakshman, and Shalender Bhasin, "The Physiological and Pharmacological Basis for the Ergogenic Effects of Androgens in Elite Sports," *Asian Journal of Andrology* 10 (2008): 351–363; Bhasin as quoted in Singer, "Does," n.p.

6. 5 News, "Usain Bolt: A Woman Would Beat Me over 800m," YouTube, September 19, 2013, https://www.youtube.com/watch?v=veSqmr-HIWs; Peter Larsson, "Track and Field All-Time Performances Homepage," http://www.alltime -athletics.com/w_800ok.htm, accessed June 1, 2018; Dominique Eisold, "International Age Records: The Best Performances by 5- to 19-Year-Old Athletes from 51 Countries," June 10, 2019, http://age-records.125mb.com

7. Izquierdo et al., "Maximal."

8. Izquierdo et al., "Maximal."

9. Shalender Bhasin, Rajan Singh, Ravi Jasuja, and Thomas W. Storer, "Androgen Effects on the Skeletal Muscle," in *Osteoporosis in Men: The Effects of Gender on Skeletal Health,* 2nd ed., ed. Eric S. Orwoll, John P. Bilezikian, and Dirk Vanderschueren (Amsterdam: Elsevier Science, 2010), 335.

10. Shalender Bhasin, Thomas Storer, Nancy Berman, et al., "The Effects of Supraphysiologic Doses of Testosterone on Muscle Size and Strength in Normal Men," *New England Journal of Medicine* 335 (1996): 1–7.

11. Lee T. Gettler, Sonny S. Agustin, and Christopher W. Kuzawa, "Testosterone, Physical Activity, and Somatic Outcomes among Filipino Males," *American*

Journal of Physical Anthropology 142 (2010): 590–599; for other data on relationship between T and fat mass in non-Western populations, see Peter T. Ellison and Catherine Panter-Brick, "Salivary Testosterone Levels among Tamang and Kami Males of Central Nepal," *Human Biology* 68 (1996): 955–965; Benjamin Campbell, Mary T. O'Rourke, and Susan F. Lipson, "Salivary Testosterone and Body Composition among Ariaal Males," *American Journal of Human Biology* 15 (2003): 697–708; Benjamin Campbell and Michael Mbizo, "Reproductive Maturation, Somatic Growth and Testosterone among Zimbabwe Boys," *Annals of Human Biology* 33 (2006): 17–25; Gettler et al., "Testosterone" 596.

12. Corey Kilgannon, "Meet 'Supergirl,' the World's Strongest Teenager," *New York Times,* December 1, 2017; Powerlifter 62, "11 Year Old Naomi Kutin Breaks the All-Time Raw 97 lb. Squat Record (again) 6-23-2013," YouTube, June 23, 2013, https://www.youtube.com/watch?v=42sDTk8hfkQ.

13. Deena Yellin, "Young Powerlifter's Challenges, Successes Are Subject of Film," NorthJersey.com, April 29, 2017, http://www.northjersey.com/story/news/bergen/fair-lawn/2017/04/29/young-powerlifters-challenges-successes-subject-film/100016258.

14. USA Powerlifting (USAPL) database, http://usapl.liftingdatabase.com/competitions-view?id=1635.

15. Crewther et al., "The Effect."

16. Jonathan P. Folland, Tracy M. McCauley, Cherry Phypers, Beth Hanson, and Sarabjit S. Mastana, "The Relationship of Testosterone and AR CAG Repeat Genotype with Knee Extensor Muscle Function of Young and Older Men," *Experimental Gerontology* 47 (2012): 437–443; Antti Mero, Laura Jaakkola, and Paavo Komi, "Serum Hormones and Physical Performance Capacity in Young Boy Athletes during a 1-Year Training Period," *European Journal of Applied Physiology and Occupational Physiology* 60 (1990): 32–37; Lone Hansen, Jens Bangsbo, Jos Twisk, and Klaus Klausen, "Development of Muscle Strength in Relation to Training Level and Testosterone in Young Male Soccer Players," *Journal of Applied Physiology* 87 (1999): 1141–1147; Alexandre Moreira, Arnaldo Mortatti, Marcelo Aoki, et al., "Role of Free Testosterone in Interpreting Physical Performance in Elite Young Brazilian Soccer Players," *Pediatric Exercise Science* 25 (2013): 186–197.

17. Paulo Gentil, James Steele, Maria C. Pereira, et al., "Comparison of Upper Body Strength Gains between Men and Women after 10 Weeks of Resistance Training," *Peerj* 4 (2016): e1627.

18. Crewther et al., "The Effect," 585.

19. Fawzi Kadi, Patrik Bonnerud, Anders Eriksson, and Lars-Eric Thornell, "The Expression of Androgen Receptors in Human Neck and Limb Muscles: Effects of Training and Self-Administration of Androgenic-Anabolic Steroids," *Histochemistry and Cell Biology* 113 (2000): 25–29.

20. As an example of this work, see Juha P. Ahtiainen, Arto Pakarinen, Markku Alen, William J. Kraemer, and Keijo Häkkinen, "Muscle Hypertrophy, Hormonal Adaptations during Strength Training in Strength-Trained and Strength Development and Untrained Men," *European Journal of Applied Physiology* 89 (2003): 555–563.

21. Grace Huang, Shehzad Basaria, Thomas G. Travison, et al., "Testosterone Dose-Response Relationships in Hysterectomized Women with or without Oophorectomy: Effects on Sexual Function, Body Composition, Muscle Performance and Physical Function in a Randomized Trial," *Menopause: The Journal of the North American Menopause Society* 21 (2014): 612–623, 619; Bhasin et al., "Androgen," 338.

22. Helen Bateup, Alan Booth, Elizabeth Shirtcliff, and Douglas Granger, "Testosterone, Cortisol and Women's Competition," *Evolution and Human Behavior* 23 (2002): 181–192; David Edwards and J. Laurel O'Neal, "Oral Contraceptives Decrease Saliva Testosterone but Do Not Affect the Rise in Testosterone Associated with Athletic Competition," *Hormones and Behavior* 56 (2009): 195–198.

23. Kevin McCaul, Brian Gladue, and Margaret Joppa, "Winning, Losing, Mood, and Testosterone," *Hormones and Behavior* 26 (1992): 486–504; Tania Oliveira, Maria Gouveia, and Rui Oliveira, "Testosterone Responsiveness to Winning and Losing Experiences in Female Soccer Players," *Psychoneuroendocrinology* 34 (2009): 1056–1064.

24. Justin Carré and Nathan Olmstead, "Social Neuroendocrinology of Human Aggression: Examining the Role of Competition-Induced Testosterone Dynamics," *Neuroscience* 286 (2015): 171–186; Shawn N. Geniole, Brian M. Bird, Erika L. Ruddick, and Justin M. Carré, "Effects of Competition Outcome on Testosterone Concentrations in Humans: An Updated Meta-Analysis," *Hormones and Behavior* 92 (2017): 37–50; Jakob L. Vingren, William J. Kraemer, Nicholas A. Ratamess, et al., "Testosterone Physiology in Resistance Exercise and Training the Up-Stream Regulatory Elements," *Sports Medicine* 40 (2010): 1037–1053.

25. Geniole et al., "Effects."

26. Andrea Henry, Jason R. Sattizahn, Greg J. Norman, Sian L. Beilock, and Dario Maestripieri, "Performance during Competition and Competition Outcome in Relation to Testosterone and Cortisol among Women," *Hormones and Behavior* 92 (2017): 82–92; Elizabeth A. Shirtcliff, Andrew R. Dismukes, Kristine P. Marceau, et al., "A Dual-Axis Approach to Understanding Neuroendocrine Development," *Developmental Psychobiology* 57 (2015): 643–653.

27. Branimir B. Radosavljevic, Milos P. Zarkovic, Svetlana D. Ignjatovic, Marijana M. Dajak, and Neda L. J. Milinkovic, "Biological Aspects of Salivary Hormones in Male Half-Marathon Performance," *Archives of Biological Sciences* 68 (2016): 495.

28. Blair T. Crewther, in conversation with the authors, August 4, 2012; Blair T. Crewther and Christian J. Cook, "Effects of Different Post-Match Recovery Interventions on Subsequent Athlete Hormonal State and Game Performance," *Physiology and Behavior* 106 (2012): 471–475.

29. Crewther et al., "Baseline."

30. G. Gregory Haff, Janna Jackson, Naoki Kawamori, et al., "Force-Time Curve Characteristics and Hormonal Alterations during an Eleven-Week Training Period in Elite Women Weightlifters," *Journal of Strength and Conditioning Research* 22 (2008): 433–446; Rosalba Gatti and Elio F. De Palo, "An Update: Salivary Hormones and Physical Exercise," *Scandinavian Journal of Medicine and Science in Sports* 21 (2011): 157–169; Lawrence D. Hayes, Fergal M. Grace,

Julien S. Baker, and Nicholas Sculthorpe, "Exercise-Induced Responses in Salivary Testosterone, Cortisol, and Their Ratios in Men: A Meta-Analysis," *Sports Medicine* 45 (2015): 713–726.

31. Helen E. Longino, "Knowledge for What? Monist, Pluralist, Pragmatist Approaches to the Sciences of Behavior," in *Philosophy of Behavioral Biology,* ed. Kathryn S. Plaisance and Thomas A. C. Reydon (Dordrecht: Springer, 2012), 25–40.

32. IAAF, "IAAF Regulations Governing Eligibility of Females with Hyperandrogenism to Compete in Women's Competitions," 2011; IOC, "Regulations on Female Hyperandrogenism," 2012/2014.

33. Katrina Karkazis and Rebecca Jordan-Young, "Debating a Testosterone 'Sex Gap,'" *Science* 348 (2015): 858–860; Stéphane Bermon, Martin Ritzén, Angelica Hirschberg, and Thomas Murray, "Are the New Policies on Hyperandrogenism in Elite Female Athletes Really out of Bounds? Response to 'Out of Bounds? A Critique of the New Policies on Hyperandrogenism in Elite Female Athletes,'" *American Journal of Bioethics* 13 (2013): 63–65; IOC, "Regulations"; Joe Simpson, Arne Ljungqvist, Malcolm Ferguson-Smith, et al., "Gender Verification in the Olympics," *JAMA* 284 (2000): 1568–1569.

34. Marie-Louise Healy, James Gibney, Claire Pentecost, Michael J. Wheeler, and Peter Sönksen, "Endocrine Profiles in 693 Elite Athletes in the Postcompetition Setting," *Clinical Endocrinology* 81(2014): 294–305.

35. Stéphane Bermon, Pierre-Yves Garnier, Angelica Lindén Hirschberg, et al., "Serum Androgen Levels in Elite Female Athletes," *Journal of Clinical Endocrinology and Metabolism* 99 (2014): 4328–4335.

36. Martin Ritzén, Arne Ljungqvist, Richard Budgett, et al., "The Regulations about Eligibility for Women with Hyperandrogenism to Compete in Women's Category Are Well Founded. A Rebuttal to the Conclusions by Healy et al.," *Clinical Endocrinology* 82 (2015): 307–308; Bermon et al., "Serum."

37. Anne Fausto-Sterling, *Sexing the Body: Gender Politics and the Construction of Sexuality* (New York: Basic Books, 2000).

38. Bermon et al., "Serum"; Rebecca Jordan-Young, Peter Sönksen, and Katrina Karkazis, "Sex, Health, and Athletes," *British Medical Journal* 348 (2014): g2926.

39. Bermon et al., "Are," 64.

40. Marco Cardinale and Michael H. Stone, "Is Testosterone Influencing Explosive Performance?," *Journal of Strength and Conditioning Research* 20 (2006): 103–107.

41. Cardinale and Stone, "Is Testosterone."

42. CAS, *CAS2014/A/3759 Dutee Chand v. Athletics Federation of India (AFI) & The International Association of Athletics Federations (IAAF),* Court of Arbitration for Sport, Lausanne, 2015, http://www.tas-cas.org/fileadmin/user_upload/award_internet.pdf.

43. Crewther et al., "Steroid Hormones," 585; Blair T. Crewther, email message to the authors, May 21, 2018.

44. Joanna Harper, "Race Times for Transgender Athletes," *Journal of Sporting Cultures and Identities* 6 (2015): 1–9, 4; CAS, *Chand v. AFI & IAAF,* 97.

45. CAS, *Chand v. AFI & IAAF,* 96; Harper, "Race Times," 5.

46. Harper, "Race Times," 2.

47. Martin Ritzén on "No Games for Women with 'Too Much' Testosterone," *The Stream,* Al Jazeera Radio, September 3, 2014, https://www.youtube.com/watch?v=5mdJfZH6BQg; Werner W. Franke and Brigitte Berendonk, "Hormonal Doping and Androgenization of Athletes: A Secret Program of the German Democratic Republic Government," *Clinical Chemistry* 43 (1997): 1269; Brigitte Berendonk, *Doping Dokumente: Von der Forschung zum Betrug* (Berline: Springer-Verlag, 1991).

48. Faryal Mirza, conversation with the authors, December 1, 2014.

49. CAS, *Chand v. AFI & IAAF,* 45, 47.

50. CAS, *Chand v. AFI & IAAF,* 54; Mark A. Sader, Kristine C. Y. McGrath, Michelle D. Hill, et al., "Androgen Receptor Gene Expression in Leucocytes Is Hormonally Regulated: Implications for Gender Differences in Disease Pathogenesis," *Clinical Endocrinology* 62 (2005), 62.

51. Suzanne Kessler, "The Medical Construction of Gender: Case Management of Intersexed Infants," *Signs* 16 (1990): 3–26; Katrina Karkazis, *Fixing Sex: Intersex, Medical Authority, and Lived Experience* (Durham, NC: Duke University Press, 2008); Georgiann Davis, *Contesting Intersex: The Dubious Diagnosis* (New York: NYU Press, 2015).

52. Katrina Karkazis, Rebecca M. Jordan-Young, Georgiann Davis, and Silvia Camporesi, "Out of Bounds? A Critique of the New Policies on Hyperandrogenism in Elite Female Athletes," *American Journal of Bioethics* 12 (2012): 3–16.

53. Bermon et al., "Serum," 4334; Bermon et al., "Are," 63, 64, emphasis added.

54. Stéphane Bermon and Pierre-Yves Garnier, "Serum Androgen Levels and Their Relation to Performance in Track and Field: Mass Spectrometry Results from 2127 Observations in Male and Female Elite Athletes," *British Journal of Sports Medicine* 51 (2017): 1309–1314.

55. IAAF, "Levelling the Playing Field in Female Sport: New Research Published in the British Journal of Sports Medicine," 2017, https://www.iaaf.org/news/press-release/hyperandrogenism-research; Katrina Karkazis and Gideon Meyerowitz-Katz, "Why the IAAF's Latest Testosterone Study Won't Help Them at CAS," *World Sport Advocate,* 2017, http://www.cecileparkmedia.com/world-sports-advocate/hottopic.asp?id=1525; Peter Sönksen, L. Dawn Bavington, and Tan Boehning, "Hyperandrogenism Controversy in Elite Women's Sport: An Examination and Critique of Recent Evidence," *British Journal of Sports Medicine,* 2018, doi: 10.1136/bjsports-2017-098446; Amanda Menier, "Use of Event-Specific Tertiles to Analyse the Relationship between Serum Androgens and Athletic Performance in Women," *British Journal of Sports Medicine,* 2018, doi: 10.1136/bjsports-2017-098464; Simon Franklin, Jonathan Ospina Betancurt, and Silvia Camporesi, "What Statistical Data of Observational Performance Can Tell Us and What They Cannot: The Case of *Dutee Chand v. AFI & IAAF,*" *British Journal of Sports Medicine* 52 (2018): 420; Katrina Karkazis and Morgan Carpenter, "Impossible 'Choices': The Inherent Harms of Regulating Women's Testosterone in Sport," *Journal of Bioethical Inquiry* 15, no. 4 (2018): 579—587,

doi: 10.1007/s11673-018-9876-3; Jeré Longman, "Did Flawed Data Lead Track Astray in Testosterone in Women?," *New York Times,* July 12, 2018; Roger Pielke, Jr., Ross Tucker, and Erik Boye, "Scientific Integrity and the IAAF Testosterone Regulations," *International Sports Law Journal* (2019), https://doi.org/10.1007/s40318-019-00143-w.

56. IAAF, "Eligibility Regulations for the Female Classification (Athletes with Difference of Sex Development)," 2018, https://www.iaaf.org/download/download?filename=2ff4d966-f16f-4a76-b387-f4eeff6480b2.pdf; Ricardo Azziz, Enrico Carmina, ZiJiang Chen, et al., "Polycystic Ovary Syndrome," *Nature Reviews. Disease Primers* 2 (2016): 16057.

57. Doriane Lambelet Coleman, "A Victory for Female Athletes Everywhere," Quillette, May 3, 2019, https://quillette.com/2019/05/03/a-victory-for-female-athletes-everywhere/; Andrew Sullivan, "Who Should Be Allowed to Compete in Women's Sports?" *Intelligencer,* May 10, 2019, http://nymag.com/intelligencer/2019/05/andrew-sullivan-who-should-be-allowed-in-womens-sports.html; Gina Kolata, "Does Testosterone Really Give Caster Semenya an Edge on the Track?" *New York Times,* May 3, 2019; Doriane Lambelet Coleman, "Sex in Sport," *Law and Contemporary Problems* 80 (2017), 75.

58. "CAS Arbitration: Caster Semenya, Athletics South Africa (ASA) and International Association of Athletics Federations (IAAF): Decision," Court of Arbitration for Sport, Media Release, May 1, 2019, https://www.tas-cas.org/fileadmin/user_upload/Media_Release_Semenya_ASA_IAAF_decision.pdf; Dan Roan, "Athletics supremo Lord Coe," Twitter, May 2, 2019, 7:22 a.m., https://twitter.com/danroan/status/1123910512192913408.

59. Patrick Fénichel, Françoise Paris, Pascal Philibert, et al., "Molecular Diagnosis of 5 Alpha-Reductase Deficiency in 4 Elite Young Female Athletes through Hormonal Screening for Hyperandrogenism," *Journal of Clinical Endocrinology and Metabolism* 98 (2013): E1057.

60. IAAF, "Regulations," 1; IOC, "IOC Addresses Eligibility of Female Athletes with Hyperandrogenism," April 5, 2011, https://www.olympic.org/news/ioc-addresses-eligibility-of-female-athletes-with-hyperandrogenism; see also Karkazis et al., "Out"; Katrina Karkazis and Rebecca Jordan-Young, "The Harrison Bergeron Olympics," *American Journal of Bioethics* 13 (2013): 66–69; Jordan-Young et al., "Sex"; Katrina Karkazis and Rebecca Jordan-Young, "The Powers of Testosterone: Obscuring Race and Regional Bias in the Regulation of Women Athletes," *Feminist Formations* 30 (2018): 1–39.

61. For an excellent summary of recent human rights statements declarations, see Human Rights Watch, "VIII. Legal Standards Regarding Intersex Children," in *"I Want to Be Like Nature Made Me": Medically Unnecessary Surgeries on Intersex Children in the US,* July 27, 2017, https://www.hrw.org/report/2017/07/25/i-want-be-nature-made-me/medically-unnecessary-surgeries-intersex-children-us#290612.

62. Karkazis and Jordan-Young, "The Powers."

63. L. Elsas, A. Ljungqvist, Malcolm Ferguson-Smith et al., "Gender Verification of Female Athletes," *Genetics in Medicine* 2, no. 4 (2000): 249–254.

Conclusion: The Social Molecule

1. Emilia Sanabria, *Plastic Bodies: Sex Hormones and Menstrual Suppression in Brazil* (Durham, NC: Duke University Press, 2016); Paul Preciado, *Testo Junkie: Sex, Drugs, and Biopolitics in the Pharmacopornographic Era* (New York: Feminist Press, 2013); Sari van Anders, "Beyond Masculinity: Testosterone, Gender/Sex, and Human Social Behavior in a Comparative Context," *Frontiers in Neuroendocrinology* 34 (2013): 198–210.

2. Karen Barad, *Meeting the Universe Halfway: Quantum Physics and the Entanglement of Matter and Meaning* (Durham, NC: Duke University Press, 2007), 40.

3. Blair T. Crewther, Zbigniew Obminski, and Christian Cook, "The Effect of Steroid Hormones on the Physical Performance of Boys and Girls During an Olympic Weightlifting Competition," *Pediatric Exercise Science* 28 (2016): 580–587.

4. Kirsty A. Walters, "Role of Androgens in Normal and Pathological Ovarian Function," *Reproduction* 149 (2015): R197; Norbert Gleicher, Andrea Weghofer, and David H. Barad, "The Role of Androgens in Follicle Maturation and Ovulation Induction: Friend or Foe of Infertility Treatment?," *Reproductive Biology and Endocrinology* 9 (2011): 116.

5. Van Anders, "Beyond"; Sari M. van Anders, Katherine L. Goldey, and Patty X. Kuo, "The Steroid/Peptide Theory of Social Bonds: Integrating Testosterone and Peptide Responses for Classifying Social Behavioral Contexts," *Psychoneuroendocrinology* 36, 9 (2011): 1265–1275.

6. Donna Haraway, *The Companion Species Manifesto: Dogs, People, and Significant Otherness* (Minneapolis: University of Minnesota Press, 2016), 112; Barad, *Meeting*.

7. Amber Benezra, Joseph DeStefano, and Jeffrey I. Gordon, "Anthropology of Microbiomes," *Proceedings of the National Academy of Sciences* 109 (2012): 6378–6381; Hannah Landecker, "Metabolism, Reproduction, and the Aftermath of Categories," *S&F Online* 11.3 (2013), http://sfonline.barnard.edu/life-un-ltd-feminism-bioscience-race/metabolism-reproduction-and-the-aftermath-of-categories/; Sebastien Abrahamsson, Filippo Bertoni, Annemarie Mol, and Rebeca Ibáñez Martín, "Living with Omega-3: New Materialism and Enduring Concerns," *Environment and Planning D: Society and Space* 33 (2015): 4–19.

8. Nelly Oudshoorn, *Beyond the Natural Body: An Archeology of Sex Hormones* (New York: Routledge, 1994); Marianne van der Wijngaard, *Reinventing the Sexes: The Biomedical Construction of Femininity and Masculinity* (Bloomington: Indiana University Press, 1997); Celia Roberts, *Messengers of Sex: Hormones, Biomedicine, and Feminism* (Cambridge: Cambridge University Press, 2007); Elizabeth A. Wilson, *Gut Feminism* (Durham, NC: Duke University Press, 2015), 5.

9. Oudshoorn, *Beyond*; van den Wijngaard, *Reinventing*; Roberts, *Messengers*; Barad, *Meeting*, 74; Annemarie Mol, *The Body Multiple: Ontology in Medical Practice* (Durham, NC: Duke University Press, 2002).

10. Preciado, *Testo*.

11. Sanabria, *Plastic*, 114.

12. For data on T and sexual desire in men, see summary in van Anders, "Beyond"; also Shalender Bhasin, Linda Woodhouse, Richard Casaburi, et al., "Testosterone Dose-Response Relationships in Healthy Young Men," *American Journal of Physiology—Endocrinology and Metabolism* 281 (2001): E1172–E1181; Andrea M. Isidori, Elisa Giannetta, Daniele Gianfrilli, et al., "Effects of Testosterone on Sexual Function in Men: Results of a Meta-Analysis," *Clinical Endocrinology* 63 (2005): 381–394; Peter B. Gray, Atam B. Singh, Linda J. Woodhouse, et al., "Dose-Dependent Effects of Testosterone on Sexual Function, Mood, and Visuospatial Cognition in Older Men," *Journal of Clinical Endocrinology and Metabolism* 90, no. 7 (2005): 3838–3846. For the relationship between T and other androgens on sexual function in women, see Susan R. Davis and Glenn D. Braunstein, "Efficacy and Safety of Testosterone in the Management of Hypoactive Sexual Desire Disorder in Postmenopausal Women," *Journal of Sexual Medicine* 9 (2012): 1134-1148; Susan R. Davis, Sonia L. Davison, Susan Donath, and Robin J. Bell, "Circulating Androgen Levels and Self-Reported Sexual Function in Women," *JAMA* 294 (2005): 91–96; but compare with Sarah Wahlin-Jacobsen, Anette Tonnes Pedersen, Ellids Kristensen, et al., "Is There a Correlation Between Androgens and Sexual Desire in Women?," *Journal of Sexual Medicine* 12 (2015): 358–373.

13. Thomas Page McBee, *Amateur: A True Story about What Makes a Man* (New York: Scribner, 2018).

14. R/Ftm, "Muscle Growth on T, What's Your Experience?," Reddit, 2011, accessed December 22, 2018, https://www.reddit.com/r/ftm/comments/2js359/muscle_growth_on_t_whats_your_experience.

15. Joan W. Scott, "The Evidence of Experience," *Critical Inquiry* 17 (1991): 773–797.

16. Evelynn M. Hammonds and Rebecca M. Herzig, *The Nature of Difference: Sciences of Race in the United States from Jefferson to Genomics* (Cambridge, MA: MIT Press, 2009), 198; Roberts, *Messengers,* 134. Most recently, sociologist Brandon Kramer's dissertation traces how research on testosterone mobilizes different notions of risk for particular racialized populations, specifically in relation to racial disparities in prostate cancer. Brandon Kramer, "Biomedical Multiplicities: How the Testosterone Industry Reconstructs Risk to Promote Pharmaceuticals," Ph.D. diss., Rutgers University, 2019.

17. Ruha Benjamin, "Innovating Inequity: If Race Is a Technology, Postracialism Is the Genius Bar," *Ethnic and Racial Studies* 39 (2016): 2227.

18. Avery F. Gordon, *Ghostly Matters: Haunting and the Sociological Imagination* (Minneapolis: University of Minnesota Press, 1997), 7; Amade M'charek, Katharina Schramm, and David Skinner, "Technologies of Belonging: The Absent Presence of Race in Europe," *Science, Technology, and Human Values* 39 (2014): 462.

19. Allan Mazur and Alan Booth, "Testosterone and Dominance in Men," *Behavioral and Brain Sciences* 21 (1998): 353–363; Allan Mazur, "Testosterone Is

High among Young Black Men with Little Education," *Frontiers in Sociology* 1 (2016): 1–5.

20. Hammonds and Herzig, *Nature,* 198.

21. Adriana Petryna, *Life Exposed: Biological Citizens after Chernobyl* (Princeton, NJ: Princeton University Press, 2002); Nikolas Rose and Carlos Novas, "Biological Citizenship," *Blackwell Companion to Global Anthropology,* ed. Aihwa Ong and Stephen J. Collier (Oxford: Blackwell, 2003).

22. Recent analyses of research and popular discourse on oxytocin are especially useful in this regard; see, for example, Fabíola Rohden and Fernanda Vecchi Alzuguir, "Unveiling Sexes, Producing Genders: The Promotion of Scientific Discoveries of Oxytocin," *Cadernos Pagu* 48 (2016); Sigrid Schmitz and Grit Höppner, "Neurofeminism and Feminist Neurosciences: A Critical Review of Contemporary Brain Research," *Frontiers in Human Neuroscience* 8 (July 2014): 546.

23. Roberts, *Messengers,* 199.

24. Anne Fausto-Sterling," The Bare Bones of Race," *Social Studies of Science* 38 (2008): 657–694; Anne Fausto-Sterling, "The Bare Bones of Sex: Part 1—Sex and Gender," *Signs* 30 (2005): 1491–1527; Amber Benezra, "Microethnicities," paper presented at the Annual Meeting of the Society for the Social Study of Science, Boston, September 1, 2017; Fausto-Sterling, *Sexing,* 255; van Anders, *Beyond.*

ACKNOWLEDGMENTS

In 2012, when we were interviewed on the BBC about sports regulations that set limits on women's natural testosterone levels, the interviewer asked us multiple times in under ten minutes why testosterone wasn't a good way to divide the male and female categories in sports. The first time she asked the question it made sense, because testosterone had been framed as jet fuel for athletes by those who formulated the regulations. But by the third time she asked us, we knew we were up against a more formidable problem: how this molecule is understood. While our interest in this hormone dates back two decades, that interview was critical in pointing us toward this project.

We are grateful to have had incredible institutional support for this work, including from the National Science Foundation, the Guggenheim Foundation, the American Council of Learned Societies, the Brocher Foundation, and a Barnard College Presidential Research Award. Many individuals supported our applications with advice, feedback, and/or recommendations, and for that help we thank Jesse Prinz, Rayna Rapp, Anne Fausto-Sterling, Bruce Kidd, Sarah Richardson, Helen Longino, Lynn Garafola, Deborah Valenze, Lisa Jean Moore, Rona Vail, and Cindy Broholm.

Over the years many research assistants have helped us, including Rebecca Dorfman, Maya Wind, Yoav Vardy, and Emily Vasquez. Liz Carlin and Brandon Kramer, who worked with us for two years and assisted with data collection, analysis, and the most fun project meetings ever, were especially central to the project.

We thank the many people who have shepherded this book at Harvard University Press, including Elizabeth Knoll, Thomas LeBien, Kathi Drummy, Janice Audet, Emeralde Jensen-Roberts, and Louise Robbins. A special shout-out to our energetic publicist, Megan Posco. Thanks also to Mary Ribesky, who managed the production. We are also grateful to the two anonymous reviewers who gave us much to chew on and unquestionably made this a better book.

We've benefited from expert editing by Frances Key Phillips, and graphic artists Sheila Goloborotko and Isabelle Lewis patiently and generously created the images.

We interviewed dozens of people during the course of this project, and we are grateful to all who so generously lent their time and expertise. We are especially grateful to Dwyn Harben, Sari van Anders, Blair Crewther, Lee Gettler, and Andrew Arnold.

KATRINA: We began this book during what felt like a simpler time. It has been shadowed by so much pain it feels like a miracle it is here at all. Sheer determination and the love and support of many made this possible. To all those who have been a part of the book and my getting here through cancer again and all else, I am extraordinarily grateful.

My family: Jackie Karkazis, Sharon Lazaneo, and the many who helped when health problems leveled us, most especially Elizabeth Robinson. Lauren Banister, my heart daughter, has been a source of tremendous joy in my life. Thank you for letting me love you. Every day I am grateful for the domestication of the wolf and my incredible shar-pei mix, Abbey Lincoln.

For all the love, conversations, texts, walks, check-ins, meals, visits to museums, music outings, and so much more, I am heartful to Løchlann Jain, Payoshni Mitra, Sarah Khan, Henry Drewal, Al Letson, Jennet Nazzal, Lizzie Reis, Ben Carrington, Laura Mamo, Helen Fitzsimmons, Jennifer Fishman, Elena Muldoon, Ali Miller, Kriz Bell, Arlene Baratz, Elise Giancola, Julia Weber, Noël Schoenleber-Fontán, Ricky Carter, Yoshi Kato, Nevin Caple, Walter Thompson, Isolde Brielmaier, Ross Wiseman, Miriam Ticktin, Roxanne Varzi, Luis Flores Canseco, and Amy Siskind. I know there are yet others who have made so much possible—if I have missed anyone here it is due to a faltering of the brain, not the heart. Huge thanks to Linda Gaal, Gabe Back-Gaal, and Bodhi for providing me with such a soft landing in New York.

Two people beloved to me, my grandma, Millie Cuyler, and Cedric Robinson, died during the time we were writing this book. I miss them terribly.

Carole Vance has been a remarkable mentor over the years, and I count myself lucky beyond belief to still benefit from her friendship and intellectual rigor.

This book stemmed in part from work on "sex testing" regulations that connected me to a world of academics and advocates doing incredible work and who have been a source of support, insight, energy, and humor: Bruce Kidd, James Bunting, Carlos Sayao, Patrick Bracher, Greg Nott, P.J. Vazel, Jennifer Doyle, Anne Lieberman, Sarah Axelson, Alison Carlson, Myron Genel, Chase Strangio, Kyle Knight, Madeleine Pape, Michelle Moore, Zhan Chaim, Eszter Kismodi, Geeta Misra, Cynthia Rothschild, Madhumita Das, Shohini Ghosh, the amazing women of the *Burn It All Down* podcast, Jovan Mircetic, Peter Sonksön, and so many more. The courage, wisdom, and determination of Dutee Chand and Caster Semenya have inspired me in ways they will never know.

For even longer, there has been the amazing community of intersex advocates who have given me the gift of alliance: Kimberly Zeiselman, Al Ittelson, Anne Tamar-Mattis, Bo Laurent, Georgiann Davis, Morgan Carpenter, and Hida Viloria, among so, so many more. Thank you.

Maria Streshinsky of *Wired* and Matt Seaton of the *New York Review of Books* are both wonderful humans and brilliant editors who provided publishing homes for words and ideas that did and didn't make it into the book, and I'm grateful to them for that.

Thanks to the Center for Biomedical Ethics at Stanford for giving me time and then even more time to complete this, and to the Honors Academy at Brooklyn College, and most especially Lisa Schwebel, for providing an incredibly supportive base from which to finish it.

One of the miracles of a book is how it can bring people closer, and in this regard I have been incredibly lucky. While studying at Columbia's School of Public Health in the 1990s, I met Frances Key Phillips. After years of a dormant friendship, she reached out and mentioned that she was an editor. She is not only an outstanding editor but a superb friend, and that this book brought us back into daily contact is a source of shared pleasure, laughter, and snark. One day I hope to make enough money to keep her on retainer, and I will happily share her so others can experience her deft edits.

Years ago, while studying for my qualifying exams, I read Suzanne Kessler and Wendy McKenna's acknowledgments for *Gender: An Ethnomethodological Approach* and was struck by what they had to say about their collaboration: "This book is truly a collaboration. The order of authorship is alphabetical, and were it not for the constraints of linear reality, the authors would be listed and indexed circularly. Any attempt to determine which parts of the book should be attributed to which author would be futile." Isolated and miserable, I longed for a similar experience. I am forever grateful to Beck for allowing me to experience the consummate collaboration I read about so long ago. I am keenly aware of the genius Beck has brought to our work and count myself among the most fortunate to have worked so closely with her for what has now been nearly a decade. There's no question we've argued, but we've also laughed so much and so hard, and if we had better filmmaking skills we would've made the hysterical videos we imagined to accompany this work. Sorry, folks. Maybe one day.

Of course, for almost all of these years, Sally Cooper, Beck's beloved partner and my dear friend, was there too. Sometimes writing in the margins of drafts, sometimes yelling from the other room with an extraordinary insight or words of encouragement, sometimes distracting us with stories of the cosmos and plants. She was the best storyteller. And she was always our first reader. I'm not sure this book will ever feel finished knowing she hasn't read it. I have texted her about our progress and will email her the final version.

BECK: It's hard to figure out precisely when this book started, but the seed of collaboration with Katrina was born in December 2011, when she contacted me about new regulations limiting natural testosterone levels in women athletes. I agreed to check out some of the scientific claims advanced by policymakers, and before I knew it had joined as a coauthor on the paper she was writing. That article snowballed into a project examining the science and ethics of those regulations, with Katrina doing amazing advocacy for athletes alongside our scholarly work. That project then blossomed into a deeper investigation of T. As we have worked side by side, through video chats and shared internet documents, while walking our dogs or biking or traveling, and every which way except with green eggs and ham, I have loved working with this woman. Katrina's excellent brain, politics, commitment to clear process, and sense of humor keep me on my toes. During our residency at the Brocher Foundation, the other fellows surely must have thought we weren't actually working, because we laughed so loud and so often. And while the work itself has been an equal collaboration, Katrina has done more than her fair share to keep this train on the tracks even when she herself was facing brutal challenges on all fronts. All respect, gratitude, and love to you, KK.

When Frances Key Phillips agreed to work as our editor, we struck gold. Thank you, Frances, for your sparkle with words and enthusiasm for the project.

While working on this book, I lost the two people dearest to my heart. First, my beloved mother, Verlie J. (Pinkey) Young, who fought her failing body and bossy children with southern charm and plenty of vinegar. My siblings, especially my sisters Nancy and Susan and my brother Tom, made it possible for me to help care for Mama, while they did the heavy lifting on the home front. Susan, Stan, John, Mark, Tom, Saphronia, and Nancy: I am proud of how we pulled together when it mattered, and I'll never forget how we sang her out of this realm.

In March 2018, after the most dazzling twenty-five-year love affair and not nearly enough trips around the sun, my precious Sally C. Cooper died. Through five years of brutal cancer treatments, Sal maintained her brilliance, generosity, fierce love, silliness, and sometimes infuriating "bluebird of happiness" outlook. I miss her with every breath and can't believe she is not here to hold the book that she supported so fiercely.

An army of love has kept me mostly upright through these losses and the general ebb and flow of life. Dozens of friends and family members have done so much for Sal and me during these years, and I ask your forgiveness in advance for not being able to name you all, but some of the most stalwart include Kate Stafford, Carolyn Patierno, and Lillian Patierno Stafford; Kim Gilmore; Cindy Broholm, Rona Vail, and Hannah Broholm-Vail; Sarah Garrison, Jane Bedell, and Derek and Anna Garrison-Bedell; Linda Gaal and Gabriel and Isaiah Back-Gaal; Rachel and Ari Efron; Amanda Joseph and John Sutton; Nancy Pfromm; Stephanie and John Proellochs; Stephanie Dowling; Phil, Jo, Rosie, Henry, and Ruben Teverow; Lizzy Schmidt, Susan Messina, Irene Lambrou, Lauren Liss, Monica Schoch-Spana, and Eleanor Bell; Miriam Ticktin; Patrick Dodd; Betsy Esch; Svati Shah; Noelle Leonard and Jennifer Brown; Lisa Gaughran; Laurie Arbeiter and Jennifer Hobbs; Ivy Arce-Kwan, and Alex, Atom, and Ahimsa; David Radoo; Janice Adharsingh; Janet Jakobsen; Christina Crosby; Elizabeth Castelli; Jeanne Stellman; Neferti Tadiar, Jon Beller, and Luna Beller-Tadiar; Marysol Asencio; Margaret Hynes; Dan Daley; Moe Angelos; Naomi Braine; Rachel Levitsky; Cynthia Rothschild; Barbara Schulman; Sheila Goloborotko; Alma Largey; Malinda Ray Allen; Errol Davis; Chadon Charles; Judith Helfand; Claudia Bloom and Graeme Evans; Raffaella Rumiata; Aishya Ali; Theo van der Meer; Geertje Mak and Ineke van Gelder; Amade M'charek; Elisa Gores; Timea Szell; Hope Dector; Jennifer McLean. Love and thanks, too, to my whole family made of Youngs, Newtons, Jordans, Dowlings, Coopers, Kimballs, and Katrina Santana.

It is impossible for me to separate my intellectual and social networks. At Barnard College, my colleagues are not just brilliant and impressive people, but so many of them are also good friends. For making me smarter and helping me in countless ways, I thank especially the Women's, Gender, and Sexuality Studies Department (Elizabeth Bernstein, Tina Campt, Janet Jakobsen, Neferti Tadiar, Manijeh Moradian, and Alex Pittman); my colleagues in the Center for Critical Interdisciplinary Studies and the Barnard Center for Research on Women, especially Yvette Christianse, Monica Miller, Pam Cobrin, Mark Nomadiou, Michelle Rowland, Kim Hall, Manu Vimilassery, Hope Dector, and Pamela Phillips; also Nancy Worman, Laura Kay, Jenna Freedman, and Lesley Sharp. I deeply appreciate the warm support from Provost Linda Bell. The extraordinary Laura Ciolkowski did far more than her share when our coteaching was interrupted by my caregiving over and over again. I'm also grateful for the camaraderie of colleagues at Columbia's Institute for Research on Women and Gender, Center for Science and Society, and Center for the Study of Social Difference, especially Lila Abu-Lughod, Pamela Smith, Roz Morris, and Katherine Franke. My former and current students are a constant source of learning, inspiration, and support. I don't dare to begin naming names, but it has been my privilege to work with you.

Colleagues and friends outside of Barnard have helped me develop ideas about sex, gender, sexuality, race, class, biology, and more: Geertje Mak, Theo van der Meer, Miriam Ticktin, Anne Fausto-Sterling, Amade M'charek, Gayle Rubin, Jennifer Terry, Stefan Dudink, David Valentine, Amber Hollibaugh, Paisley Currah, Sarah Richardson, Sahar Sadjadi, Kerwin Kaye, Helena Hansen, and others in the "Symbioses" group (especially Dorothy Roberts), and the Neurogenderings Network (especially Anelis Kaiser, Cordelia Fine, Giordana Grossi, and Gina Rippon). Carole S. Vance has been the sort of mentor and friend that I only dream of being, though I pledge to keep trying.

Maisey Cooper Jordan-Young has made sure that I continue to get up and get outside every morning, play on a daily basis, and don't forget to eat.

Publication Credits

The Introduction builds on ideas first discussed in K. Karkazis, "The Testosterone Myth," *Wired*, April 2018, and K. Karkazis, "The Masculine Mystique of T," *New York Review of Books Daily*, June 28, 2018. Chapter 7 incorporates excerpts that first appeared in the following publications: K. Karkazis and R. Jordan-Young, "Debating a Testosterone 'Sex Gap,'" *Science* 348, no. 6236 (2015): 858–860, reprinted with permission from AAAS; K. Karkazis, "Stop Talking about Testosterone— There's No Such Thing as a 'True Sex,'" *Guardian*, March 6, 2019, reprinted courtesy of Guardian News & Media Ltd; and R. Jordan-Young and K. Karkazis, "Four Myths about Testosterone," *Observations* (blog), *Scientific American*, June 18, 2019, https://blogs.scientificamerican.com/observations/4-myths-about-testosterone/. In addition, some themes explored in Chapter 7 were initially developed in the following articles: K. Karkazis, R. Jordan-Young, G. Davis, and S. Camporesi, "Out of Bounds? A Critique of Policies on Hyperandrogenism in Elite Female Athletes," *American Journal of Bioethics* 12, no. 7 (2012): 3–16; K. Karkazis and R. Jordan-Young, "The Powers of Testosterone: Obscuring Race and Regional Bias in the Regulation of Women Athletes," *Feminist Formations* 30, no. 2 (2018): 1–39; and K. Karkazis and R. Jordan-Young, "The Myth of Testosterone," Opinion, *The New York Times*, May 3, 2019, https://www.nytimes.com/2019/05/03/opinion/testosterone-caster-semenya.html.

INDEX

Page numbers followed by *f* or *n* indicate figures and notes.